ML

ACPL ITEM
DISCARDED

## DO NOT REMOVE
## CARDS FROM POCKET

## ALLEN COUNTY PUBLIC LIBRARY

## FORT WAYNE, INDIANA 46802

You may return this book to any agency, branch,
or bookmobile of the Allen County Public Library.

DEMCO

ENERGY RESOURCES AND POLICIES OF THE
MIDDLE EAST AND NORTH AFRICA

---

# AN APPRAISAL OF
# OPEC OIL POLICIES

---

ALI. M. JAIDAH

LONGMAN
LONDON AND NEW YORK

Longman Group Limited,
*Longman House, Burnt Mill, Harlow,*
*Essex CM20 2JE, England*
*and Associated Companies throughout the world.*

© Longman Group Limited 1983

First published *1983*

ISBN 0 582 78361 5

**British Library Cataloguing in Publication Data**

Jaidah, Ali M.
An appraisal of OPEC oil policies. — (Energy resources
and policies of the Middle East and North Africa)
1. Organization of the Petroleum Exporting Countries
I. Title     II. Series
338.2'3        HD9560.1.066

ISBN 0-582-78361-5

**Library of Congress Cataloging in Publication Data**

Jaidah, Ali M., 1941–
An appraisal of OPEC oil policies.
(Energy resources and policies of the Middle East and
North Africa)
Includes index.
1. Petroleum industry and trade — Government policy —
Addresses, essays, lectures. 2. Petroleum products —
Price policy — Addresses, essays, lectures.
3. Organization of Petroleum Exporting Countries —
Addresses, essays, lectures. I. Title. II. Title: Appraisal of
OPEC oil policies. III. Series.
HD9560.6.J34 1983     382.'4228'0601     83–5450
ISBN 0-582-78361-5

Set in 11/12pt Bembo AM 4560
Printed in Great Britain
by Spottiswoode Ballantyne Ltd, Colchester and London

# CONTENTS

# INTRODUCTION

The papers collected in this book were written over a period of six years. Some were addresses made at various international conferences when I was the Secretary-General of OPEC 1977–78; and the others after I returned to Qatar to take up my current position as Managing Director of the Qatar General Petroleum Corporation. All these papers reflect my personal views as they have evolved over the years and do not engage in any way the responsibility of OPEC, the Qatar General Petroleum Corporation or my own Government.

I have been involved with oil and OPEC affairs since the mid-1960s in various official capacities in my own country and as one of its representatives to OPEC committees and boards, and on occasions as a member of the Qatari delegation to OPEC Conferences of Oil Ministers. The subject matter of this book has been and still is the daily concern of my whole professional life.

In these capacities I have been most privileged to witness the major oil events which have established OPEC as an international force and a powerful instrument of change on the world political and economic scene. There were crises and difficult moments when a slack market or an imbalance in the bargaining strength of oil companies and governments created difficulties for OPEC; but I also witnessed the momentous days of 1973 when OPEC achieved the oil price revolution, and when member countries retrieved their full sovereignty over their main national asset – their petroleum resources.

*Changes and Continuities in the Oil World*

In the past twenty years the oil scene has been in a constant state of flux. In the 1960s, demand for and supplies of oil grew at a high annual rate, yet the oil price remained virtually constant in dollar terms and the revenue per barrel accruing to oil-producing countries declined in real terms. OPEC however made some modest progress in the ten years following its establishment in 1960. It inhibited companies from lowering the posted price of oil as happened on several occasions in the late 1950s. OPEC obtained from the companies better fiscal arrangements (expensing of royalties, higher tax rates) for its member countries. It began to claim and to obtain equity participation for governments in the forms operating oil concessions in member countries.

In the early 1970s the petroleum market became tight and a long-term energy crisis was predicted by many observers of the oil industry. It is in this climate that the OPEC price revolution took place in October–December 1973. This revolution changed the system of oil price determination which had prevailed for long decades. Oil companies used to set the posted price unilaterally and negotiate tax rates with the producing countries when circumstances forced them to do so. The events of 1973 inaugurated a new era in which OPEC set the oil price autonomously without recourse to negotiations and bargaining with the companies. Other important changes followed transforming the features of the oil scene. Government participation was increased to 100 per cent – which simply means that Governments took back the concessions after paying a compensation for the fixed assets of the operating companies – in most of the member countries. The National Oil Companies of producing countries expanded their marketing function selling directly larger proportions of the country's oil output to a wide range of foreign buyers. All decisions relating to prices, terms of contracts, production levels, exploration and development of petroleum resources came under the exclusive authority of the government.

The oil price revolution of 1973 did not however mark the beginning of an era of continued market tightness. The conditions of the oil market vary significantly from time to time. The tight market of 1973–74 was followed by a period of relative slackness in 1975 mainly due to a world economic recession. Many observers then asked whether OPEC will be able to retain its recent and hard-won achievements. Western economists and politicians having labelled OPEC as a cartel believed that OPEC will collapse like all cartels as soon as market conditions become unfavourable. They overlooked the fact that OPEC was established when the economic conditions facing member countries were particularly adverse, and that its raison d'etre was precisely to defend the interests of these members when the interests were being threatened. It has survived before in an unfavourable environment, largely due to OPEC's ability to adapt swiftly to new realities. Why should OPEC collapse at a time when the motivation for solidarity was at its strongest?

The oil market was slack once again in 1977–78, and the same doubt about OPEC's ability to hold the price line were expressed with renewed emphasis. But this episode was not followed by a collapse but by an unexpected and dramatic event – the Iranian Revolution. The political disruption in Iran caused uncertainties about the security of oil supplies, and caused a scramble for oil. OPEC followed the strong movement which pushed prices up; but, contrary to common views, it did not lead the increases and did not raise its price to the maximum recorded on the spot market at the height of the crisis. But there was an aftermath.

The oil price structure became distorted. Official sales prices of similar varieties of crude oil were set at different levels by different member countries. For a time, OPEC had two reference prices for oil – an actual price for Saudi Arabian Light 34° API and a higher "deemed" marker price on the basis of which other OPEC members fixed the value of their own crudes. This disorderly price structure was not perceived as a serious issue when the market was tight, but it rendered OPEC members vulnerable to considerable pressures from companies and other buyers when oil demand began to decline in 1981. Since then, and at the beginning of 1983, OPEC faced the most serious challenge to its controlling power over oil prices. The combination of a severe drop in oil consumption (largely but not exclusively due to economic recession) and of a severe inventory drawdown, reduced total demand for oil. Because OPEC is cast in the role of the "residual supplier", the reduction in aggregate oil demand was transmitted in its entirety to OPEC. Member countries suffer the whole brunt of any demand shortfall. Similarly, when oil demand grows, OPEC countries are expected by the rest of the world to provide all the incremental supplies. This is an unsatisfactory state of affairs to be in. Sharp declines in oil output and oil revenues play havoc with development plans of individual countries and put the weakest members under considerable strain. Sharp increases in output can cause problems of a different nature. They encourage governments to make spending commitments which may not be sustainable when market conditions turn round. More seriously, they prevent the adoption of any consistent depletion policy that may be designed to preserve the long-term economic interest of the country.

Despite these continual changes in oil market conditions and in the world economic environment in which OPEC operates, historical developments in different periods have certain similar features and strong elements of discontinuity manifest themselves. These continuities seem to underly all the apparent changes. Thus, as noted in Chapter 11:

> "The oil market continues to function whether the supply-demand balance is slack or tight. OPEC remains a strong institution with a long experience of crises and successes. The oil-producing countries face today, as they did yesterday, the same long-term problems of economic development and of excessive reliance on a depletable resource. The oil-consuming countries on one day seem concerned about the security of oil supplies and on another totally forgetful of the significance of the issues. Yet the threat of a long-term crunch never fades away".

Put differently, the point is that short-term changes in market conditions, however conspicuous and dramatic in their immediate impact, should not conceal the fundamental fact that oil is a depletable resource. The depletion of a finite asset poses long-term problems for both producing and consuming countries. When a short-term glut arises, the perception of these problems tends to become blurred; when a shortage occurs the perception becomes sharper and may sometimes provoke a sense of crisis. In reality, the problems are always there. The time-horizon over which they are likely to manifest themselves may sketch away or come closer. But the true nature of these problems and their real significance do not change in the same way.

The issue for producing countries — those which belong to the Third World — is how to transform their depletable resource, which is often their only source of economic wealth, into the reproducible assets of economic development. The development horizon is longer in most cases than the expected economic life of oil reserves. This is a serious and complex issue which does not fade away when current market conditions change in one direction or another. This theme runs throughout the various papers included in this book.

The issue for consuming countries is to prepare a smooth transition from the oil era to another energy era. The exhaustibility of oil resources means that this transition is inevitable. Sooner or later the problem will have to be faced. If the transition is not carefully prepared and planned in a way that minimises its economic and social costs, disturbing disruptions could arise. True, industrialised economies may be able to adjust in the end and to survive, in one way or another, a disruptive crisis. But the costs of such an adjustment could be crippling. The main difficulty is that investment in energy involves very long lead-in times. Costs and gestation lags mean that the timing of investment decisions must be carefully chosen. To do too much, too early is extremely costly and wasteful. To do too little, too late will also involve expensive disruptions.

## The Objectives

These long-term issues which survive short-term upheavals in markets impose an enduring objective for the international community: to organise the energy transition in a way which enables the producing countries to secure the means of their economic development and the consuming countries to move away from one energy era to another at minimal costs, that is without an unnecessary waste of scarce resources.

If we were to translate their objective from OPEC's point of view it would involve a responsibility on member countries to pursue their long-term economic interests in a manner consistent with an international obligation to contribute to the orderly transition into the post-oil era.

Historically, OPEC's primary goal has been to recover the legitimate right of a producing country to administer in a sovereign manner the price of its natural resources.

"Since its birth in 1960, our Organisation has been characterized by a clarity of objectives. It envisaged a central objective to which it diligently stuck for more than ten years. That objective revolved around the concept that oil prices are an integral part of the basic national interests of member countries and that their decisions to change them are their legitimate right ...". (Chapter 4)

This primary goal was indeed achieved in 1973 and nobody seriously challenges today a country's prerogative to set "the prices of its oil exports, in the same way as other exporters of finished commodities exercise the right to set the prices of their own exports ..." (Chapter 4). What the consuming world refuses to accept is the right of these producing countries to exercise their prerogatives collectively and to administer the price of oil in international trade. Spurious theoretical arguments (mentioned in Chapters 1 and 8) are made against this system of price administration. To say that prices of an exhaustible resource should relate to marginal costs or to sing the merits of perfectly competitive markets miss all the relevant points. The pricing of non-reproducible resources must reflect a relationship with the costs of their long-term substitutes or their growing scarcity in the long-run. And perfect competition in the absence of a chain of futures markets does not ensure efficient pricing of a depletable resource. As somebody put it very aptly, to say that the price of oil is its production cost means that the price of an asset should be equal to the cost involved in its liquidation. These pseudo-theoretical arguments made against OPEC pricing neither reflect good theory nor an understanding of the factors which have actually influenced the determination of oil prices throughout the history of the industry.

It is not sufficient however to recover the prerogative to determine prices. The problems of how to administer prices, in a way which would enable OPEC to maximize benefits in the long-run and to provide correct signals for the allocation of resources in the world economy, have to be faced. These are complex problems and OPEC has naturally met difficulties as well as successes in dealing with them.

Further, the price goal is itself part of a much wider objective. What is then the ultimate aim? It is indeed to solve the long-term development problem already mentioned here above.

"The export of our finite petroleum resources is directly connected with the process of our economic transformation from countries which export a single primary commodity to countries which are capable of self-sustained economic growth, a perspective which must be linked in its time horizon with a period within which our oil exports will diminish or cease forever". (Chapter 4)

This central goal is far from being fulfilled. Part of the difficulty arises from the attitudes of the industrialized countries and from policies which limit the scope of fruitful and mutually beneficial cooperation. Part of the difficulty is due to inherent weaknesses of

OPEC which stem from the fact that its member countries are all part of the Third World. They are still constrained — despite the oil wealth — by the many problems of economic under-development.

Yet as emphasized in both this introduction and the book, OPEC believes that it has an even wider role to play in contributing to an international solution of the long-term energy problem. This role can build on OPEC strengths. It involves a national administration of the price of oil, the main instrument at our disposal for signalling the true state of oil scarcity. It calls for practical policy measures applied at both the national and the international level, measures which can only succeed in a climate of true cooperation, freed from inconsiderate or truly hostile attitudes.

Prices; relationships with the industrial and the developing world; strengths and weaknesses of OPEC in connection with the objectives it seeks to achieve. These are the concepts which I have tried to link together in this introductory analysis. These are also the main themes which run through the pages of this book. Having placed these themes in context, I propose now to explore briefly their contents and implications in order to draw a few relevant conclusions for OPEC and for the international economic order of which it is an integral part.

## Prices

Prices perform two roles. They provide signals for the allocation of resources in the economy and they determine the distribution of income. The second function of prices concerns the *national* economic interests of OPEC members, because the price of oil, is for its greatest part, the revenue which accrues to the oil-producing country per barrel of export. The first function concerns OPEC in its *international* role as administrator of the price of oil.

Before 1973, oil-producing countries did not exercise the controlling power over prices. Price administration was in the hands of the major petroleum companies; and the responsibility to signal to the world the increasing scarcity of vital but depletable oil was entirely theirs. But

"The companies which were vested with the 'controlling power' over oil prices were not concerned with long-term scarcity. They were more involved with the short-term conditions of the oil market and were engaged actively in expanding their market outlets. Long-term considerations called for price increases, but short-term interests and considerations resulted instead in the lowering of the real price of oil". (Chapter 10)

For OPEC the first priority was to obtain price increases or for changes which will satisfy the revenue objectives of its members. It should be recalled that the price of oil is the predominant source of revenue for the developing oil-exporting countries. It is indeed the only source of foreign-exchange earnings. The price received directly affects the lives of all inhabitants of those countries. Their livelihood, their welfare, the prospects of long-term economic development, all critically depend on it. (Chapter 8).

OPEC members could not accept the situation prevailing in the 1960s when their revenues per barrel were less than one dollar while petroleum products fetched the equivalent of 12–14 dollars per barrel, the difference being made for a large part of excise taxes levied by the consuming countries' governments.

OPEC fought over a long period to obtain its legitimate share of the oil rent which should accrue to the owner of the resources rather than to the governments of importing countries which neither have property rights nor contribute in any way to the development of oil reserves. Companies have a right to normal profits related to the productive effort put in to the risks taken. In some special circumstances the final consumers may conceivably be allowed some proportion of the rent. But there is no case for governments of importing countries to appropriate any part of it.

After OPEC regained the right to determine prices it still faced the problems of maintaining the real purchasing power of its unit revenues. As it is well known that the purchasing power is affected by inflation and by exchange rate fluctuations. Thus, between 1974 and 1979 OPEC members found that the real value of their revenues was being substantially eroded by very high rates of inflation and by a depreciation of the dollar. As real values are eroded, the incentive to restore their levels by raising the nominal price of oil becomes very strong. The price revolutions of the 1970s have taken place against a background of frustration and resentment about the sharp decline in real revenues.

Oil price administration involves both the reference price and the set of price differentials for the many varieties of crude. The manner in which the reference price is increased is of considerable importance because sudden rises can have disruptive economic effects, while long periods of price stagnation allow strains to develop and lead in the end to unwanted price explosions. Of equal importance is the need to keep the whole oil price structure constantly in equilibrium which means that differentials should correctly reflect the relative values of different crudes.

The lessons drawn from the two episodes of price rises in 1973 and in 1979/80 have clearly demonstrated that sudden shocks have detrimental effects on oil-consuming and oil-producing countries alike. But the 1973 shock was largely due to the imbalances created by a long period of price stagnation. It was the failure of major oil companies to raise prices in the 1960s in a manner reflecting the pressures of growing demand on the supplies of a depletable resource which made it necessary to adjust prices through a very large increment. And the price stagnation of 1974–79 played also a role in the subsequent price shock. As the real prices of oil were falling in these years, companies began to reduce their inventories and were caught during the 1979 crisis without an adequate cushion. As noted earlier, producing countries were frustrated by the erosion of their revenues and were eager to secure a new adjustment as soon as market conditions allowed.

There is no doubt, as mentioned in Chapter 10 that a price determination system allowing for smaller regular increases in the price of oil would contribute to avoiding shocks, or at least to making them less severe. The OPEC Long Term Strategy document prepared in 1978–79 included a proposal for a price formula designed to achieve steady increases. The Strategy has not yet been adopted as revision of the work done became necessary because of change in circumstances.

The issue of price differentials is treated in several papers in this collection (Chapters 8, 10 and 12). The price structure tends to become distorted when the market is tight. In this situation, every producer is a price maker, and may take advantage of his bargaining power to obtain the most favourable terms available. He need not worry about a reduction in sales if his prices are higher than those of other producers because buyers are willing to take full volumes from everywhere. Naturally, prices tend to get out of line. This happened in 1973/74 and again in 1979/80.

But gluts follow shortages; and oil producers often find themselves entering the period of glut with a distorted price structure inherited from the preceding phase. Their ability to cope with the glut is weakened when the price structure is disorderly. In a buyers' market companies have room for manoeuvre. They can switch away from the high-price producer and secure additional supplies from the cheaper sources. Some producers take the full brunt of a reduction in demand while others escape unscathed. Strains tend to develop within OPEC.

The advantages of maintaining an orderly price structure cannot be sufficiently emphasized. There is no doubt in my mind that OPEC would have had a much easier ride during the "oil demand crisis" of 1981–83 if this essential requirement of good price administration had been fulfilled, and some member countries did not indulge in price discounting.

## Relationships with Industrial and Developing Countries

Oil is a vital commodity, the life blood of the modern industrial world. Its economic importance is considerable, not only because oil accounts for a large proportion of trade flows, but because the costs of disruption should oil supplies become insecure or short are extremely high. By virtue of its function as administrator of oil prices, OPEC has acquired an international role, a certain amount of power and responsibilities. As an organisation of sovereign states it finds itself in an interface with the rest of the international community — industrial and developing nations alike.

The interests of various groups of nations naturally diverge. To take the case of oil, it is evident that buyers would prefer the lowest possible price and sellers the highest. It is not surprising that importing and exporting countries perceive their immediate economic interests in different ways. But there is a mutuality of interests on another and higher plane. All parties have much to gain from an orderly transition to the non-oil era.

The high price, which buyers resent, often represents an allocative signal which induces greater efficiency in the use of energy, bigger investments for alternative energy supplies, a larger research and development effort. All these are required at some stage if the energy transition is to take place. These points are mentioned in Chapters 3 and 10. From this point of view there is no necessary conflict between the immediate interests of producers in terms of revenues and the long-term interests of consuming countries in terms of their economic future.

Attempts to fulfil the profound aspiration for economic development of OPEC countries are not only beneficial to them but involve gains for the rest of the world. Through development OPEC countries become expanding markets for the exports of industrial and Third World countries. Industrialization in OPEC countries could improve the pattern of comparative advantages in the world economy and increase international trade. Development also fosters labour and capital flows across borders, bringing the skills where they are best rewarded and investment funds where they can earn higher returns.

Yet the emergence of OPEC was not generally greeted as an opportunity for greater international cooperation to the mutual benefits of all members of the international community. The first reaction was fairly negative.

"... the West ... was unwilling to accept the fait accompli that there had been a shift in international economic relations towards the developing countries. It was perhaps too revolutionary for them and they feared that such action by a group of Third World countries would serve as an example for producers of other vital commodities to follow". (Chapter 7)

There seems to be a reluctance — especially in the US — to accept a degree of economic reliance on developing countries for a commodity like oil. But this is an interdependent world. The OPEC countries are themselves overwhelmingly dependent on the industrialized nations for their vital subsistence needs (food and consumer goods) and for the investment goods required for economic development.

There were attempts to combat OPEC's role as price administrator. The International Energy Agency was established to create, initially at least, a strong front against OPEC. But there were more subtle attempts to foster unfavourable market conditions for OPEC. Non-OPEC oil supplies were developed even where economic criteria did not justify the investment. A system developed which ensured that non-OPEC supplies are called upon first to meet demand. This puts OPEC in the role of a 'residual supplier' bearing the full brunt of any variation in world demand. Stock-piling was allowed to take place in a disruptive cyclical way. Inventories are drawn down when demand for oil is slack and the effect of such behaviour is to make a weak market even weaker (see Chapters 3, 6 and 7).

OPEC does not receive the full support required for the economic development of its members. As a "residual supplier" it suffers large and unanticipated fluctuations in revenues. Protectionism in industrialized countries hinders the expansion of non-oil exports from OPEC countries and from the rest of the Third World. Projects which would be profitable if they enjoyed some access to export markets cannot be undertaken because this access is barred. Finally, there are serious obstacles to the transfer of technology, a theme which appears in several places in this book.

The industrialized nations are guilty of short-sightedness which manifests itself in a variety of areas. Take natural gas, for example. For a long time the economic value of gas was ignored and huge amounts were flared in OPEC countries' oilfields. Gas had no immediate economic value to the concessionaire companies in the 1950s and 1960s, yet a significant long-term value for the countries themselves. Conservation measures were called for, but the producing countries could not afford them and the companies simply turned a blind eye.

In another area, short-sightedness was also apparent. The industrialized countries did their utmost to prevent the North-South dialogue from taking off the ground. They resisted successfully at the CIEC in Paris attempts to alter the international economic order in favour of the Third World. OPEC countries being themselves developing countries have endeavoured to enhance solidarity with the Third World by extending aid on a very large scale and by promoting the concepts of a new international economic order. But the strategy of the industrialized world is to drive a wedge between oil-exporting and oil-importing developing countries. They wanted to isolate energy from all the other issues on the agenda. A deal on energy alone would have meant gains for the rich industrial world without any compensatory measures in favour of the Third World. The need for better terms of trade for a wide variety of primary commodities, for transfer of technology, for more concessional aid and for the removal of protectionist barriers which hinder exports from developing countries is paramount. To ignore these issues does not really benefit the industrialized world since everybody will gain from the enrichment of the poor through greater and better opportunities for investment and exchange.

## OPEC's Strengths and Weaknesses

OPEC has shown its strength when it faced adverse conditions. Despite predictions to the contrary, it maintained the solidarity of its members throughout the 1960s, in 1975, and in 1977–78. This strength arises from a deep understanding of the value of OPEC's objectives. The gains from sticking together are substantial; the losses that would be incurred after a collapse are simply enormous. Twenty years of common experience, of learning through crises and difficulties the rules of the gain have cemented the organisation.

OPEC countries make valuable contributions to the economies of the industrialized world. Large financial surpluses are placed in the banking system of the West; and OPEC countries have become large markets for exports of goods and services and offer considerable opportunities for profitable business and construction work. Thus, OPEC is not deprived of economic leverage. Yet OPEC countries have too often shyed away from exerting this leverage for the promotion of causes close to their hearts, the most important being the economic causes of the Third World.

Much remains to be done to preserve and enhance the strength of OPEC. The organisation remains fairly slow in its response to rapid market changes. Its share in the world oil output is shrinking. Of immediate concern OPEC since 1981 faces difficulties which threaten its cohesion and its ability to defend the price level. In response to low level of demand for OPEC oil some of its member countries along with other non-OPEC producing countries retain their level of production by indulging in various forms of price cuttings. No serious attempt was even made to define and implement a production policy necessary to underpin prices in a slack market and to ensure the success of any long-term pricing strategy. The administration of price structure sometimes left much to be desired. And oil-producing countries have not attempted yet to rationalize and coordinate their development plans in ways which will avoid duplication and improve access to foreign skills and advanced technology.

The domestic energy policies of most oil-producing countries have been defective. Thus petroleum products are usually sold very cheaply within producing countries on the grounds that the population should benefit from the country's main resource. The danger of this policy is that it encourages very rapid and very wasteful growth of oil consumption. Some OPEC countries will soon exhaust their resources in this way and find themselves becoming net importers of energy within a short time-horizon.

*Conclusions*

Two main conclusions may be drawn from this analysis. The urgent task which OPEC faces is to build on its strength and remedy its weaknesses. This is necessary to enhance solidarity within the organisation and to enable OPEC to face successfully major challenges and obstacles which may lie ahead. What is needed in this respect has been already mentioned in this introduction and repeatedly emphasized in this book.

The programme of action includes an oil production policy. OPEC adopted such a programme for the first time in March 1982, and in the period starting May/June 1982 some member countries began to flout the production agreement and to discredit both the output programme and the agreed price structure. The need for such a step was argued in most of my papers and the arguments in favour have become familiar. More remains to be done on price administration, most urgently eliminating all forms of discounts and on the preparation for quick adoption of a long-term strategy. OPEC could profitably do more work on the problem of industrial development downstream and encourage coordination of activities between national oil companies.

The second task is to promote a spirit of global interdependence. In this area, relationships between OPEC and the rest of the Third World have a privileged place. Despite the difficulties raised by the Western countries and their protracted resistance there is scope for developing the relationship with industrialized countries. These relationships will only be improved when the terms of exchange become more equal. At present, oil is exchanged for dollars and consumers complain about a price which may seem nominally high but which fails to give producing countries real value for their resources. Oil should produce development. And the most fruitful area of cooperation is indeed economic development.

Any endeavour to fulfil these two tasks will contribute to the resolution of the problems posed by the depletability of oil. These are major problems of transition: a development transition for the producing countries and an energy transition for the world at large. The two problems are related. These are long-term issues which transcend volatile changes in market conditions. They represent a considerable challenge for the international community. This book attempts to define and explore the nature of this challenge. I would think the effort undertaken here worthwhile if I succeeded in alerting the reader to the urgency of the task.

February 1983                                    Ali M. Jaidah

CHAPTER 1

# PRICING OF OIL:

# ROLE OF THE CONTROLLING POWER

*The following is the substantial text of a lecture presented at an Energy Seminar sponsored by Harvard University, Cambridge, Massachussetts on 9 May, 1977.*

## 1. Introduction

It is a welcome opportunity for me to be able to address an academic audience like you. There are several challenges in this setting. Many of you will, no doubt, assume positions from which you can influence the policies of the United States in various degrees. If I can arouse your curiosity in the problems I shall raise, this in itself will be a worthwhile contribution to the debate on the desired New Economic Order we are involved in shaping.

The developing countries of the Third World, as well as the advanced countries of OECD, are committed, in principle, to reform the present inadequate state of affairs. A particularly interesting aspect of this encounter is that we might be starting our dialogue from very distant positions. You are citizens of the most advanced and influential industrial society, while I speak for a group of developing countries blamed for so many of the world's troubles: recession, inflation, energy shortages, currency problems ... etc.

There is nothing more fatal than mistrust and misunderstanding. In our world, so precariously surviving on a complexity of delicate balances, policies based on false assumption might undermine the whole balance. A great deal of suffering and unhappiness could be averted by keeping a dialogue going on between all parties concerned. Hopefully, a dialogue would bring about positive cooperation; but the least it could do would be to defuse unwarranted confrontation.

For these reasons, I have picked an area which might provoke most debate: oil prices. In my belief, facing the issue is better than ducking it.

There is a great body of literature on oil prices. Distinguished economists and oil men have approached the subject from various aspects. Many authors have tried to explain the nature of oil prices and their evolution. Several of them tackled the question from the point of view of the Majors or the consuming countries. Others tried hard to be objective.

Perhaps the most thorough work in the field of petroleum economics is the research by Professor Adelman during the sixties and culminating in the publication of his masterly volume *"The World Petroleum Market"*[1] round about the time when all his predictions were shattered by events proving the exact opposite of what he preached. During the sixties, he showed, with painstaking thoroughness, that the marginal cost of supplying the world with oil for 15–20 years was about 20 cents per barrel, or at most, 30 cents per barrel. His conclusion was that oil prices had only one way to go: down.

## 2. Controlling Power as Price Maker

It is, indeed, a source of disappointment to realize that all these worthy men gave us little insight into the nature of oil prices and their probable direction. It is, however, neither rewarding nor charitable to accuse all of them of misjudgment.

I would suggest that tools of analysis applied in the past, tools many of us still use, are not fitting to the situation under consideration. Equilibrium economics in a competitive environment, with its obsession to equate marginal costs with prices, is useless when we do not have a competitive situation and where the margin is nowhere in sight.

Why were oil prices around US$ 3.00/bbl before World War II? Why were they just over US$ 1.00 in the Middle East a few years ago? Why are they now over US$ 12.00? We will never find the answer in marginal cost nor in Ricardian rent. We can only find it in an objective analysis of the forces which have a vested interest in oil prices and have the power to influence those prices. Such forces can bring to bear various pressures to manipulate

[1] Morris Albert Adelman: *The World Petroleum Market*, Baltimore and London; The John Hopkins University Press, 1972.

prices to serve their own interests; whether by direct or indirect means.

It is important to remember that the oil industry, until very recently, was dominated by Western companies and Western governments, if we set aside the oil industry in the Soviet Union. If we go back to the origin of Oil Agreements in the areas which now comprise OPEC member countries, we find that those so-called Agreements were one-sided affairs. Concessions were carved up along the lines in which spheres of influence were determined as part of the spoils which followed the great wars. Many Oil Agreements, especially those concluded before the World War II were the result of negotiations between unequal parties. Many oil companies which used to operate in OPEC member countries resembled a state within a state. They were extensions of the foreign policies of the "Controlling Powers" – especially the United States and Great Britain. Other European powers were involved in the spoils, but to a lesser extent. The International Majors, which are basically Anglo-Saxon, entered our countries as the instruments of the traditional European powers on the one hand and the ascending political and economic influence of the United States, on the other hand.

These Majors exercised the "Controlling Power" on behalf of their respective patrons. This is well-documented history now. Where we are mislead is in the pretension that the Majors made their pricing policy purely on commercial terms, subject to traditional competitive equilibria.

Prior to World War II, the bulk of oil entering international trade came from the USA. Therefore, the oil exported from the Gulf of Mexico was not allowed to be undercut by cheaper oil coming from the new producers of the Middle East. This gave rise to the "Gulf–Plus" formula, which stipulated that the price of oil anywhere outside the US was equal to its price in the Gulf of Mexico plus the cost of transporting it to any point of sale. During the war effort, the European Allies were pressing for cheaper fuel. New theories were promulgated to pronounce that this indeed was an economic equilibrium. The famous "Watershed" theories first demanded that oil coming into Europe had to have a similar price whether it was exported from the Middle East or from the Gulf of Mexico. This meant bringing down the oil prices in the Middle East. Later, oil prices in the Middle East were brought down further when the same Watershed theories were adjusted to allow the United States to import oil from the Middle East. By the late forties oil prices in the Middle East were lower than those on the Eastern Seaboard of the USA by the same magnitude by which they were higher prior to the war.

What had dictated those price reductions? The producing countries were given at the time a flat four English shillings per ton. No substantial changes in the cost of production have taken place. The only change can be traced to the shift in the balance of forces making up the "Controlling Power". The higher prices forced on Europe earlier gave way to the new relationship of the Allies during and immediately after the War. Cheap fuels for Europe were an essential prerequisite for its recovery, and thus an integral part of the Marshall Plan. The oil resources of the present members of OPEC, especially those in the Middle East, were speedily developed in one of the most massive operations in recent history. Perhaps this cheap source of energy contributed more than any other factor to the rapid recovery of Europe in the Post-War period. It is not too much to ask now that Europe should repay some of its debt to the developing countries of OPEC in their own efforts at industrialization.

Until the early fifties, when "Profit-Sharing" arrangements prevailed, oil companies were not interested in the price level as long as they could maximize their total net income after taxation in their own countries. They paid no taxes to the producing countries. Their net profits were determined by the difference between their total costs and average income on sales of petroleum products. Crude oil prices were never real outside the United States. There, they represented the income of oil producers who were selling oil to the big oil companies. Those companies bought oil in the real sense of the word from the numerous private producers of Texas and Louisiana. Nobody sold oil in the Middle East – and nobody bought it in any real sense. There were only big swaps between the international Majors. Crude oil prices were mere entries in the books of the Majors.

Variations in the Posted Prices of Middle Eastern oil, prior to profit-sharing, meant little to the producing countries, apart from the fact that the value of their resources was being downgraded. But when profit-sharing came about, it was based on Posted Prices, which were already low enough. As if this were more than we deserved, Posted Prices were further lowered in the spring of 1959 by $.05 – .25 per barrel in Venezuela and by an average of $.18 per barrel in the Middle East. In August of 1960, Middle Eastern oil prices were further reduced by another $.04 – .14 per barrel. This unreasonable exercise of power by price makers was the direct provocation which forced oil producers to form an organization for oil exporters to counterbalance the prevailing controlling power of the Majors backed by the governments of the USA and Europe. OPEC came into being one month after the last price reduction.

To the Majors, payments to producing governments were a form of cost. Any elementary textbook would tell you that corporations tend to minimize average costs. With Posted Prices at $1.80 per barrel, the cost of lifting oil was less than $1.00, and the books of the companies were showing profits in the range of $.80 per barrel. This meant a rate of return on investment in the region of 100 per cent in many producing countries. The book-value of those investments was low indeed, and many investments were fully amortized. To make their exorbitant profits at the production stage look acceptable, the Majors used to lump all their Eastern Hemisphere accounts together, thereby showing more moderate returns. They reduced their apparent profitability (and tax liability) further by ploughing back a lot of investment to their home countries (see *Fig 1, page 141*).

During the sixties the companies resorted to a number of tactics to prevent the newly formed OPEC from realizing its aim of adjusting Posted Prices upwards. The arguments they advanced were mainly as follows:

1.  There was an increasing surplus of oil in the world, as attested by surplus capacity reaching 50 per cent in some cases.
2.  Soviet exports were forcing the Majors to put down their prices.
3.  Independent oil companies were creating hot competition and forcing prices down.

First, the story of surplus sounds odd for companies which were pouring billions into the North Sea, Alaska, and elsewhere outside OPEC in order to develop new capacities (see *Fig 1, page 141*). In any case, anyone who looks at the production of Kuwait, Iran and Iraq during the fifties, would realize that the companies not only turned the taps on and off to meet exactly the quantities consumed, but also that they developed surplus capacities to pressurize one country or the other. During the sixties, they were pressurizing the whole of OPEC. The truth is that spare production capacity has always been part of the oil industry without resulting in over-supply.

Secondly, Soviet oil was never a real threat. It came very slowly, moving from about 16 million tons of crude oil and petroleum products exported outside the socialist countries, to about 30 million tons in 1965. By 1970, those exports were about 48 million tons, representing less than 4 per cent of world oil exports during that year, at a time when annual demand increases were about twice as much.

Thirdly, independent oil companies never made prices in the situation that existed at the time, but rather followed the Majors in pricing decisions as marginal producers normally do. It is curious to note that those very independents were paying more in taxes and other forms of income to OPEC countries than the Majors, despite the fact that their actual costs of production were higher and their fields smaller. Many of them had to sell their crude back to the Majors because they had no outlets.

The Majors went a step further. A new phenomenon appeared in the mid-sixties called "arms-length transactions", a sort of occasional show to demonstrate how prices of crude oil were going down. We were treated to a spectacle where crude oil prices were shown to go down from $1.80 per barrel in the early sixties to about $1.30 per barrel by the end of that decade. Despite this decline which was supposed to have a detrimental effect on net corporate profits, those net profits rose from about $2.8 billion in 1961 to about $4.8 billion in 1970. (see *Fig 2, page 141*)

Here we come to the real crux of the matter. Prices of petroleum products are the only really meaningful reference, if we want to analyse the oil industry. It is perhaps here that the traditional economic analysts made their mistake. They tried to analyse crude oil prices – which were of little relevance – and ignored the real point where demand makes itself felt: the final prices of petroleum products.

A composite barrel of petroleum products in the main European markets averaged about $13.60 when OPEC came into being. If we remove the direct costs of production, transportation, refining and

distribution, we might be left with more than $10 per barrel, even
if we allow normal return on capital for investment in the various
stages. There are many variations in costs as well as in product
prices from one country to the other. We are here interested only
in the order of magnitude. What we want to stress very strongly is
that the real price for petroleum is the price for petroleum
products and not Posted Prices or what is called arms-length prices
of crude oil. It is at this price that demand interacts with supply. It
is at this price that the consumer makes his choice felt. It follows
that the split of profits at this price level reflects the real
relationship between the forces that make up the "Controlling
Power".

When OPEC came into being, the consumer governments of
Europe took over $7 per barrel – or about 70 per cent of the
profit, the producer governments getting only about $0.76, or 7.6
per cent (see *Fig 3, page 142*). The companies' books might not
show the full net income per barrel, but that is attributable to the
fact that they were using profits generated from operations in
OPEC to finance less rewarding activities.

It is food for thought to ponder on the idea that, if the
governments of the main consuming countries had been prepared
to pass down the total rent generated by oil to its rightful resource
owners, the individual consumers might be still paying for a
composite barrel of products, prices not much different from what
they used to do before OPEC was set up.

## 3. Sharing the Control of Power

OPEC succeeded in putting an end to the unilateral determination
of Posted Prices by the oil companies right from the start. It
concentrated on the only real hard fact at its end of the operation:
income per barrel exported. In this connection, it got more from
independent companies than from the Majors. Those newcomers
were prepared to share their controlling power with OPEC
countries.

New agreements were reached with those newcomers in new
areas or areas relinquished by the Majors. The new type of
agreements recognized the sovereignty of our countries over our
resources. They brought in the principle of participation with our
national companies (or the relevant state organs).

New conservation laws were enacted to put an end to the wasteful production techniques practiced by the Majors, making some use of US conservation experience. New regulations brought the old concessions gradually into the fold. OPEC income from oil exported from the Middle East moved from about $0.76 per barrel when OPEC came into being in 1960, to $1.35 in 1971, and to $1.56 in 1972. What is more important, OPEC began to share the Controlling Power.

This new status enabled OPEC to take over the decision to determine its income per barrel exported on October 16, 1973. Thus, OPEC increased its income per barrel on the Marker Crude to an average of about $9 in 1974 and to $10 in 1975.

This completely new situation caught the other side of the Controlling Power (i.e., the oil companies and the importing countries of the OECD) off guard. The taxes levied by consumer governments on a composite barrel of products in Europe were about $7.60 when OPEC came into being. They had gone up to $8.70 in 1971 and to $11.40 in 1973, which represents a much greater increase than OPEC income. OPEC income had gone up by $1.54 per barrel while consumer governments in Europe increased their share by $4.40 for the same period. But, in 1974, OPEC increased its share by $6.70 per barrel, while the European governments had to bring down their average income by about $1.00 per barrel.

This was a truly historic shift in controlling power. But it did not last long. By 1975, those governments pushed up their income by about $4.40 per barrel of products, and continue to enhance it even further at the expense of the final consumer.

These facts reveal a situation which cannot be analysed in terms of the classical theory of international trade where all parties increase their welfare when each concentrates on his comparative advantage. Here we have a situation of competition to get a larger slice of the same cake: the value generated by a barrel of crude oil. The main contending parties are:

— The consuming countries of the OECD
— The international Majors, and
— The member countries of OPEC

Since 1973, the international Majors and other companies of the OECD Countries have lost most of their claim to independent behaviour, especially since the creation of the International Energy Agency and its complex controls and logistic powers.

If we measure the share of power by the slice of the income, we find only 30 per cent of the total worth of a barrel in Europe of about $33.20 going to OPEC Countries in 1975. The consumer governments get 45 per cent of that in direct taxes. They also get further income through the Value Added Tax (VAT), and yet another slice through taxes on the income of oil companies.

If we analyse the respective share in the power to control, in terms of decision, OPEC might turn out to share even less than 40 per cent of the real decisions. The reasons for this belief are many:

1.  The OECD countries still make the real price decisions at the end of the line and, therefore, they control total demand. OPEC can only manipulate their income at the point of export.
2.  The OECD countries hold the vital element of pricing goods and services on which we spend our money. Our import price index has more than doubled since the beginning of 1974, thus halving the value of our oil exports.
3.  OPEC countries have little influence on the International Monetary and Financial Institutions which matter so much to the currency of pricing (i.e., the US Dollar) and to our temporary surpluses which are recycled to the OECD countries.
4.  The OECD countries control the basic decisions affecting the development of indigenous alternatives to oil, as well as the direction of research and development in this area.

Those who have wasted so much ink and paper to warn the world of the evils of the dictatorial powers of an irresponsible OPEC are either misinformed, or else they are unrepentant advocates of Western hegemony over the Third World.

OPEC has thrown all of its weight into the effort to bring about a New Economic Order for the world; a world in which those who had little say in the evaluation of their resources would share in the controlling power. We want a fair deal; a package in which the producers, manufacturers and users of the exhaustible resources of the world would cooperate in striving towards a higher quality of life for all people.

## 4. Conceptual Problems in Oil-Pricing

If, for a moment, we imagine that the world is one economy, without boundaries and without conflicting interests, and then ask the following question: How to price petroleum resources? What answers do we get?

Liquid and gaseous hydrocarbons (oil and gas) are finite in their availability. They are exhaustible resources like so many of the natural resources on which our civilization was built. We cannot use up those resources as if they were inexhaustible.

The economic theory of exhaustible resources tells us that the optimal price for such finite resources – like the prices of goods held in a warehouse – should rise annually in real terms at rates equal to interest rates or some appropriate discount factor for time. This is necessary to induce an orderly production profile over time and to maximize the present value of the profit or rent derived from the reserves of our common resources. This price curve which rises over time should help to restrict demand, the closer we come to the point of scarcity and depletion. In this way, other technologies would meet future human requirements at the higher price level, a price level sufficient to make the necessary technologies viable. The price curve should not be so steep that alternatives take over before the resource has been used up, nor so flat that the result is depletion without making alternatives viable. The conditions for such optimal pricing and production policy can never be met in real life, but must be kept in mind.

One of the difficulties about petroleum prices is that hydrocarbons are not a homogeneous commodity meeting a single demand. Residual fuel oil for power generation is the least valuable form of using hydrocarbons in direct burning. A more efficient form of consuming petroleum is in the internal combustion engine. Perhaps the most rewarding use of hydrocarbons is not in the field of energy, but in using them as a petrochemical feed to produce synthetics, fertilizers, proteins, etc. Each of these uses represents an optimal price level different from the others. We should not think of a price level for oil which takes only power generation into account and depletes the resource on that basis alone.

Another vital issue has to be examined in our theoretical unified world. Future generations cannot be doomed by our present waste of scarce resources. Some consideration for intergenerational equity has to be made. Theoretical price curves are not enough. We might find out that direct conservation is called for, not only to avoid waste (like US conservation and pro-rationing laws), but also to provide for the future. This deliberate holding back of production would necessarily mean upward price adjustment. We have to give up a little to save the future.

Now we have to come back to reality and see what all this means in the harsher light of our divided world:

The pricing of petroleum products, we find, tends to encourage wasteful burning and heavily penalize the more efficient users. As an example of this, we find that in Italy, a power station can buy fuel at about $91.20 per ton at a time when other alternatives in power generation are estimated to cost over $150 per ton of oil equivalent. Premium gasoline, on the other hand, costs the motorist over $790 per ton, while alternatives are nowhere in sight. This is a general pattern in all OECD countries, including the US. This sort of situation is only possible because prices of crude oil are separated from prices of products.

By this separation, two things are achieved: less is given to producers, and some of the high rentability of petroleum resources is used to subsidize inefficient uses. If we cannot ask for oil prices to go up the level of the most efficient use (with the highest rent), at least we can initially argue that oil prices could be related to the alternatives of the least efficient use, which is, in this case, nuclear energy.

Rent and rentability can only be related to the resource base from which they arise. The fact that this rent goes largely to consumer governments has to be gradually redressed. One aspect has to be eliminated as speedily as possible for the sake of rational utilization of fossil fuels. Much of the rent derived from oil is used by the industrial countries in direct and indirect subsidies to coal and nuclear industries. To find out the real cost of nuclear energy, you have to let electricity prices rise gradually to the point where private capital starts pouring into nuclear power plants. All environmental and waste-disposal costs have to be added to the investment requirements. Only at that level could we say that the cost of alternative technology is real. This level can serve as a floor for oil prices – not a ceiling. We must start from now to think about limiting the use of oil in direct burning and to divert it to more efficient uses. Those uses naturally have greater rentability.

## 5. Some Interim Considerations

While striving to find a workable formula for international cooperation in the UN, UNCTAD, the Paris Conference and other fora, OPEC is concerned about the implications of crude oil pricing in the interim period.

OPEC has accepted the responsibility of meeting world requirements for crude oil. This has resulted in production rates in some countries in excess of their immediate requirements for foreign exchange. Temporary monetary surpluses have been generated. To avoid any disturbance in the already shaky international money markets, those surpluses were recycled to the OECD countries. Large sums were also set aside for the needy countries of the Third World. These surpluses must be given profitable and secure investment channels by the consumers whose high demand requirements continue to force the generation of such surpluses.

A more serious problem is the effect of inflation on oil prices and prices of imports of the OPEC countries. Unless inflation is checked, OPEC has no choice but to adjust its oil prices to maintain the real value of its exports. Long price freezes mean a decline in real oil prices. If this continues, we are bound to face another crisis coupled with absolute energy shortages. A solution to the high levels of inflation cannot be found in lowering the prices of imported oil. It is to be found in a better control of money supply and a wage policy related to productivity. We do not accept the argument that the poorer nations of the world, including OPEC countries which are all developing countries, should pay the price for keeping up the living standard in the advanced countries beyond the means of those countries.

As I said earlier, OPEC countries are developing countries. The money we get from the export of oil is only a means to achieve our development requirements. We have to transform our exhaustible assets into fixed capital formation which enables our economies to achieve sustained growth. In this development effort, we need a counter-commitment on the part of the consumers to help in making such development possible, especially with respect to the transfer of technology at reasonable terms.

Coupled with this development objective, we have to stress another responsibility to the future generation of our countries. We should not be expected to deplete our energy resources without consideration for the energy requirements of our own countries in the future. It would be a great folly indeed, if we should allow our oil to be exported at low prices and thus condemn our future generations to import energy at much higher costs. Either we are provided with a global energy package whereby we are guaranteed alternative technologies at reasonable costs, or we will have to modify gradually our open-ended commitment to meet the world's demand for oil.

(Annex of Tables to Chapter 1 may be found on pps. 141, 142).

CHAPTER 2

# THE PRESENT AND FUTURE ROLE

# OF THE NATIONAL OIL COMPANIES

*The following is the substantial text of an address delivered at the OPEC Seminar at Vienna, 10–12 October, 1977.*

It is with great pleasure that I avail myself of this opportunity to welcome such a distinguished gathering to OPEC's Headquarters.

May I, at the outset, make it quite clear that although the theme of this Seminar is stated as being *"The Present and Future Role of the National Oil Companies"*, especially those of OPEC member countries, it is desirable that its scope should be extended to encompass the diverse areas pertinent to the roles of other parties engaged in the world petroleum industry.

If anything can be regarded as characteristic of OPEC since its inception, it has been the clarity with which the Organization has formulated and pursued its objectives and the carefully thought-out institutional framework for achieving their realization. These are two important lessons we learned right at the beginning. It was always our aim to take over the decision-making function in the setting of oil prices: This process can be said to have begun in earnest with the Tehran Agreement of 1971 and to have been completed in October 1973, when OPEC, for the first time in its history, unilaterally adjusted the prices for its crude oil. The second task which the Organization had set itself was, ultimately, to have complete control of the hydrocarbon industry in its sovereign territories. This devolution of power is a gradual process which is still continuing, but at this point in time it can be said that OPEC has already advanced far along the road towards achievement of this goal.

It is unnecessary for me, in this address, to go into the history of these accomplishments; it has, after all, been said before, and for those represented here today, who were themselves principal actors in this particular stage of events, a recapitulation would be superfluous. No, it is not the purpose of this Seminar to deal with the past; but in considering action for the future one can, and sometimes should, draw on past experience. I wish, therefore, to refer again to the two lessons to which I made reference earlier.

OPEC, as we have seen, has achieved legal ownership of the production stage; our national oil companies and government agencies have been entrusted with the management of the upstream stage of the oil industry. But the transfer of legal ownership is not enough in itself. We cannot stop there; we cannot simply rest content to be mere producers of natural resources, with no further role to play.

Standing, as we now are, at a cross-roads in our history, we must remember those two important lessons and apply them to the new situation. Here again, as at the beginning, we, through OPEC, must present a united front in formulating and putting forward our legitimate claims, since only through our continuing solidarity can we hope to repeat the successes of earlier years.

As I see it, these new objectives can be formulated thus:

1) The efficient management of the oil industry by nationals of OPEC member countries at all levels;

2) The development of an indigenous technological base, backed by domestic research institutions, capable of contributing increasingly to the needs of the oil section in member countries;

3) The speedy transformation of the role of OPEC member countries from that of raw material exporters to manufacturers by carrying out certain downstream operations, especially with regard to refining and petrochemicals. In this way the national oil industry should become the central pivot in the process of industrialization.

Obviously, the tasks referred to above cannot be carried out overnight. Neither can they be achieved without the understanding and cooperation of other parties involved in the international oil industry.

It is my hope, therefore, that in the course of this Seminar we shall be able to evolve a clearer definition of the tasks that lie ahead as well as to determine the most suitable institutional framework for achieving those ends.

In the past years, our national oil companies (NOCs), have been faced with an immense challenge to their capabilities and resources; and it is to their credit that the production stage, over which they hold control, has not become an area of imbalances and tension. The fact remains, however, that they are faced with great problems with regard to manpower, technology and administration.

In the context of the OPEC member countries' drive for overall economic development, we cannot ignore our recent experience. We have been buying technology at exorbitant prices, whilst turn-key projects have proved to be tied to the suppliers of technology for patents, spare parts, operations, research, etc. The present terms of transfer of technology are a source of deep concern to us, not only because they are of a grudging nature, but also because we are denied free access to the markets for the products in the developed countries. The argument that the advanced consuming countries have surplus refining and petrochemical capacity is totally unacceptable to us because, every day, we hear of new plants being constructed in those very countries.

We have no other choice, therefore, but to take the risk of constructing our own plants. Surely, however, more consideration could be accorded to oil producers – at least in the fields of refining and petrochemicals. How can we talk of cooperation seriously when OPEC member countries, taken together, command a meagre 6 per cent of the world's refining capacity? And what about our negligible share of the world's petrochemical industry? A mere 3.2 per cent! Neither can one help pondering about the significance of the fact that less than 3 per cent of member countries' crude oil exports are transported by means of their own tankers.

I must say, in all seriousness, that unless greater progress is made in redressing the imbalances in this area, our member countries will have no recourse but to adopt collective strategies to achieve their aims.

Prevention is better than cure. Today all of the parties involved in determining the future of the world oil industry are represented in this room. We have, therefore, the unique opportunity to examine the issues involved and to seek ways and means of dealing with them to our mutual satisfaction. As far as NOCs of OPEC member countries are concerned, this necessitates closer and more effective cooperation.

Finally, I would just like to point out that the age of "national economies" devoted to selfish national interests can no longer be sustained: it must give way to a truly international economy and a new world economic order, in which the wealth and resources of the world will be more equitably distributed.

# CHAPTER 3

# OPEC AND THE FUTURE OIL SUPPLY

*The following is the substantial text of a lecture delivered to an Energy Conference Sponsored by the Swedish Petroleum Institute, Stockholm, Sweden on 1 November, 1977.*

The theme I have been asked to address – the Future Oil Supply as seen from the OPEC point of view – is one that has been cited at many such gatherings as that of today, and I sincerely hope that the points I am about to bring forward will, perhaps, cast a new slant on a matter of immediate concern to the world.

Since the 1800's, the chief sources of the world's industrial energy have been fossil fuels, mainly coal, and oil and gas, which are non-renewable in nature. Until 1900, the energy derived from oil and gas, as compared with that of coal, was almost negligible (amounting to 7.9 per cent). Since then the contribution of oil to the total energy supply steadily increased until, in 1968, it was approximately equal to that of coal; thereafter it increased more rapidly. If natural gas and natural gas liquids are added to the picture, the energy represented by the petroleum group of fuels would have been about 63 per cent of the total in 1972, while the share of coal, which was the prevailing source of energy in 1900 (about 89 per cent) had dwindled by the same year to 28 per cent. Furthermore, the rate of oil consumption increase was greater than the rate of growth in energy demand, while the reverse was true for coal.

When looking at the consumption figures, it can be clearly seen that although coal has been used for about 1,000 years, half of the quantities produced so far have been mined since the beginning of World War II, while half of the world's accumulated production of petroleum has occurred since 1956. In brief, most of the world's consumption of energy from fossil fuels has taken place within the last thirty-five years.

From a review of world coal production from 1860 until the present day, it becomes evident that production was increasing at an exponential rate of about 4 per cent except for the period between the two World Wars when the rate was 0.75 per cent annually; whereas oil production was marked by a steady exponential increase of 6.9 per cent per annum, interrupted only by the Depression and World War II.

The reason why oil and gas resources came to occupy such a unique place in the supply of energy undoubtedly lies in the fact that they were regarded – until very recently – as cheap, easily accessible sources of energy, and unique in their scope of utility.

After 1973, the attention of the world was focused, as never before, on the problems related to natural resources and energy. Both the new strategies of OPEC countries and the dramatic warning of the Club of Rome, awakened the world to the notion of physical limitation, inducing governments, private institutions and others, to give deeper thought to the availability and future supply of natural resources.

In reviewing the various literature and estimates on the future availability of hydrocarbon resources, one immediately recognizes two schools of thought, namely:

a)   The *pessimistic* – based on the idea that due to the modern industrial rates of consumption, the time span for exhaustion of the bulk of the various fossil fuels is very limited. It has frequently been stated that the declining discovery rate of hydrocarbons is a consequence of the fact that the more the resource base is discovered, the less there remains to be discovered. Consequently, during the period of human exploitation the resources or fossil fuels may be considered to consist of fixed initial supplies which are continually being diminished by human consumption. The world as a whole should be generally concerned over the disappearance of these exhaustible assets and should devise logical means whereby to regulate their exploitation. The ultimate objective would be to attain an optimum distribution of these resources' use over time, in order not to exhaust them before alternatives become available and to ensure a smooth transitional period from an economy based exclusively on non-renewable fossil fuel resources into an economy based on renewable resources. By all accounts, the duration of this transition will be measured in decades rather than centuries; however, all efforts should be made to prolong this as far as possible.

b) The *optimistic* – whereby it is stated that the fear that petroleum may become scarce is probably unfounded when looking into the future and considering the large variety of natural petroleum sources available. Present published estimates of petroleum resources refer in a practical term to a very tiny fraction of the total nature-made petroleum base, namely the production of petroleum from a conventional oil or gas field. However, the availability of petroleum resources is not dependent primarily on geological, but rather on technological and economic considerations. As petroleum prices increase, the resources become more plentiful due to the fact that those which were earlier regarded as uneconomical now become economical based on the facts that

(1) additional percentages of oil can be recovered from already known reserves by applying, at cost, enhanced recovery techniques;
(2) additional amounts of resource base can be converted into reserves by extending exploration efforts to unexplored areas; and
(3) additional types of resources, such as tar sands and oil shales, can become economical to produce and thus contribute to the amount of oil and gas which remains available.

It is then concluded that nature-made petroleum resources are large and sufficient enough to permit the world a smooth transition to alternative energy sources, provided that the time is used to the best advantage and not squandered on short-term objectives. In addition, it can be seen that the foregoing views do not widely differ from those of OPEC, namely that resources should be used more efficiently and at rational prices if a series of crises are to be avoided.

With present total world recoverable reserves standing at 659 billion barrels, and consumption at 20.8 billion barrels per year a thirty-three year supply appears to exist, provided, of course, that consumption continues at the same level and no additional reserves are discovered. However, consumption is increasing much more rapidly than are new discoveries and, therefore, it is time to direct attention to possible new sources of nature-made petroleum and gas, as well as to the improvement of recovery.

Past and present estimates of ultimate recoverable reserves of oil and gas have a rather extensive range – a reasonable consensus of opinion favours the ultimate recoverable reserves at between 1,600 and 2,000 billion, although some put the figure at 4,000 billion, based on linear extrapolation of the past trend for the increase of such reserves. Although there are some good reasons for believing that, in time, our knowledge and expertise will increase allowing additional resources to become recoverable, further continued expansion on the same scale seems unlikely, not only because there has been a great increase in the amount of exploration undertaken, but also in its geographical distribution and effectiveness.

With the estimated availability of additional conventional resources, and the variety of unconventional sources of petroleum and gas, the age of petroleum may be considerably extended, enabling it to play a more important role in the difficult transition to a new energy economy, hopefully based on renewable resources. The conditions under which such a transition will be managed are today raising momentous issues. In this connection a very important question is: how long, in fact, can or must the transitional period last, and what is the potential of being able to rely for as long as possible on oil and gas resources during this period? The ultimate aim of society must be to prolong it in order to allow for a real break-through in the development of alternative – especially renewable – sources of energy so that they can be made available to the extent required and at reasonable cost by the time non-renewable resources are nearing exhaustion.

To achieve the above objectives, it will be necessary for the policy-maker:

*.A. To increase the hydrocarbon resource base from conventional and/or unconventional resources*

Regardless of what is the accurate estimate for the ultimately recoverable amount of hydrocarbons, the extent and availability of oil and natural gas, the fuller use of conventional resources opened up by enhanced recovery, and the possibility of using some hitherto untapped unconventional resources, are all governed by the efforts made, and success achieved, in solving the various constraints which are causing significant hindrance to the rapid development of additional reserves. These constraints are to be found under the headings of:

    i)     technical and manpower
    ii)    financial and economical
    iii)   political and environmental

They are all interrelated and cannot be considered in isolation one from the other.

If one wishes to look at the future availability of petroleum at prices which it is assumed will be higher in the future than they are today, one has to take into consideration the variety of possible naturally occurring petroleum resources, how to explore for them, and how to extract the oil and gas.

In this connection, the first task is to undertake a proper assessment of the hydrocarbon resource potential, not only through research and advanced technology, but also by extending exploration to previously untouched areas. For this reason, an evaluation of the many readily accessible, potentially productive, petroleum basins of the world becomes a necessity, and here we believe OPEC countries have great potential.

A further future deterrent in the field of commercial application of highly technological resource development will be the lack of adequate proficient expertise and shortages of equipment and here particular attention should be given by both industry and universities.

The present manpower shortage is, in part, due to the economic situation and may be seen as the delayed effect of financial and political constraints while, at the same time, there is inevitably a feed-back from the manpower shortage in the ability of the industry to respond rapidly to calls for more exploration and quicker advances in engineering and technological development.

The links between technological and financial constraints are very close and, whether public, private or mixed, firms have to be able to foresee future revenue to justify present expenditure, while research of the type required for a better understanding of the new, less conventional resources is undertaken with an eye to the future. Quite apart from the indirect constraint to the effect of minimal incentives for development and research, the actual sums required for the application of existing technology are beyond the capacity of some sectors of the industry – even of some governments. One may well ask: how good are the chances of finding the much larger sums required for the exploration and development of – as yet – unidentified resources in most of the countries of the Third World, or for that matter for the exploitation of known resources? Here, governmental, regional and international entities must play a larger role so as to provide the capital needed to initiate and accelerate research, in addition to expanding exploration and development at the required rate.

The question of environmental constraint is real and requires most serious attention. Gone are the days when exploration and development could proceed without regard to environmental factors; now it is necessary to study the consequences of any policy action on the environment and to consider safety problems concerning waste and its disposal, recycling, emission, or other pollution factors, land use and conservation, the impact on wildlife, and so on. Indeed, the environmental constraint has taken on such a prominent place in the public mind that it could – on a large scale – hinder or delay the development of the hydrocarbon base, especially from unconventional resources.

Political constraints are directly interrelated with economic ones; the policy of maintaining an artificially low price for natural gas and oil has led to a halt in research and development for the distant future. Furthermore, an energy policy based on a local self-assessed interest might well hinder the rapid development of discovery and additions of new reserves, globally, whereas investment decisions will face acute difficulties without political decisions and sacrifices.

*B. To develop rapidly the alternative sources of energy, especially those of a renewable nature*

Due to the fact that prior to 1973, oil and gas prices were artificially maintained at unjustifiably low levels, which were just above the production costs in some areas, these exhaustible commodities – for some countries the only source of revenue – were undervalued for a considerable number of years and were overconsumed as if they were inexhaustible. Even following

1973–74, when OPEC adjusted its crude prices in order to receive a more adequate value for its crudes, the actual pattern of consumption of the various fossil fuels continued to follow the same trend. It is worth mentioning at this stage that OPEC's 1973 decision was aimed, in addition to deriving a more adequate value for its exports, at giving an economic incentive to the development of alternative sources of energy. However, such development has not yet assumed the momentum envisaged and it may well be that the present price of crude is still too low to give the required stimulus for this undertaking at a faster pace.

For about a quarter of a century nuclear energy has been regarded by many as the natural successor to petroleum and gas in providing clean and safe energy sources. However, the perspective of the proliferation of atomic reactors is now generating widespread and deep concern with regard to safety, while costs are increasingly being challenged. Furthermore, due to the limited utilization of solar energy, and the fact that efforts towards, and financial support of, research and development have been minimal and restricted, no real break-through is anticipated, although it is hoped that those countries with financial and technological capabilities will direct more effort in this field to enable the energy supply system to be based no longer on hydrocarbons alone, but on a multiplicity of energy sources, in particular renewable resources.

## C. To adopt sound conservation policies

Extending the transition period requires far more than development of new supplies or diversification of resources; it also calls for radical intervention from the demand side. The demand for natural petroleum is a complex issue because it is affected by prices, by the development of alternative sources of energy and by the change in end-use technology which takes place slowly. Conservation of hydrocarbon resources entails the capacity to use these resources in such a way as to avoid waste and their exhaustion before alternatives become available.

Resources are something that vary in time; they increase and develop as man learns to use processes and products, the mechanisms and qualities of which were previously unknown. Conservation, therefore, must take on a dynamic meaning: it must be linked with prices and, in some way, reference should always be made to the differing interests of both producers and consumers.

At this stage it is worth examining the various practices and measures adopted by major consumers towards the hydrocarbon conservation issue. OECD countries, while adopting sound long-term measures towards their indigenous resources, such as the restraint on production in Norway, limited well production in the United States and the exploitation of coal in Europe and America, fail to apply the same attitude and sound concept towards foreign supplies of these exhaustible resources.

The conservation measures adopted so far do not constitute a real concession to the exhaustibility of these resources. Studies carried out on the energy habits of consumer countries show the possibility not only of drastically reducing the wastage of energy, but also, more generally, of organizing energy systems different from the present ones, able to integrate a multiplicity of sources, both old and new using the material available to the fullest extent.

In view of this, major consuming developed countries should pursue a more vigorous policy aimed at:

a.  Elimination of all forms of waste stemming from irrational users of hydrocarbons.

b.  Revision of production structures and those for transforming energy sources in order to eliminate relevant existing losses in conversion and transportation.

c.  Integration in the system of such energy sources as can contribute to the energy balance, especially those from renewable sources.

A rational approach for consuming hydrocarbon resources entails a global approach based on a reasonable appreciation by consumers of the productive capacity and requirements of producers, and by the latter of the essential energy needs of the consumer countries.

This may now be the appropriate moment for me to sketch out the role of OPEC in the future supply of oil as I see it today.

It is felt necessary that the life span of OPEC's hydrocarbon reserves should be extended as long as possible, so as to provide its member countries with the necessary time to transform their depletable deposits into more permanent assets, through development and industrialization. It is worth mentioning here that after

1973, when people started talking about the drastic redistribution of resources to the advantage of our member countries, they omitted to mention the fact that these resources had not yet been transformed, except to a very small extent, into wealth. Today, some of our member countries, with increased foreign exchange at their disposal, are encountering serious difficulties in converting the value of their oil exports into real wealth. Most OPEC countries still require a long time to achieve this transformation, while those that had a greater capacity for economic growth have seen this diminish drastically as foreign trade has placed considerable constraints on their development. I believe that you will all agree that accumulating large bank deposits in a short time, quite apart from the possibility of using them to produce more income for the country concerned, corresponds neither to the best criterion of conservation, nor to the best development policy. In this regard, it is felt that the developed countries have a moral obligation to assist in accelerating the development process of our developing countries through the adequate and timely transfer of modern technology. Within this period we need to achieve not only self-sustained economic growth, independently of oil revenues, but some of our member countries have to reach a stage where their non-petroleum exports can pay for their eventual energy imports. Projecting the world-wide geographical distribution of coal, oil, gas, and other sources, it is expected that the energy source balance will soon shift from the noticeably rich areas in liquid and gaseous hydrocarbons (OPEC) to the present consuming countries, which are relatively rich in coal and have the greater possibility of developing alternative energy technologies. Therefore, in the long run, when our hydrocarbon resources are depleted, we can only satisfy our future energy requirements by importing energy.

Taking into consideration the cost, supply, and logistics of alternative sources of energy, as well as the depletable nature and versatility of oil and gas, OPEC member countries should receive, in the medium run, an equitable and remunerative price level for their exports and a stable real value of income per unit exported. In this regard, it must be said that using a non-renewable resource primarily as a fuel is an economic waste and efforts should, therefore, be devoted to enhance its usage in the non-energy sector.

In this context, and given rational pricing of gaseous and liquid petroleum products, an orderly replacement should take effect, allowing petroleum products to contribute more to transportation and non-energy uses of hydrocarbons. In our view, the best way to achieve this is by pricing crude oil at levels not less than the long-term supply of alternatives. At a later stage, and to encourage replacement of oil and gas as thermal fuels, they have to be priced above the most costly energy alternatives so as to encourage the replacement process.

Having summarized briefly OPEC's general approach to production and pricing, it might be appropriate to give you a preliminary assessment of reserves, production and exports from its member countries in the future.

The proven oil reserves within OPEC member countries now stand at about 450 billion barrels, with an estimated total amount of recoverable reserves of some 550 billion barrels. If the present production level (30 MMb/d) were to be maintained, OPEC's output would start declining after 26 years and would be depleted around the year 2025. Therefore, if the life span of total OPEC reserves is to be extended, it can only be done by either reducing the present production level and/or increasing the availability of the resource base through intensified exploration efforts, horizontally or vertically, in the highly potential basins within OPEC member countries, in addition to utilizing the most up-to-date production techniques, such as enhanced recovery, thereby increasing the amount recoverable.

Until 1973, the dominant trend of activities in petroleum exploration and development by foreign operators within OPEC member countries centred around the search for conventional – but large – petroleum deposits located in accessible areas and the use of production technology to extract free-flowing crudes. Hence a minimum of exploration was carried out by these operators, who concentrated rather on development and production from large finds, omitting to give the small ones an appropriate evaluation. This traditional approach was considered by the operation as justifiable during the period of low oil prices. They regarded the exploitation of other petroleum deposits as uneconomical and neglected other extracting technologies for petroleum resources.

The amount of proven oil reserves mentioned above is based on a recovery factor of about 30 per cent. Increasing recovery efficiency will, of course, add a substantial amount to the availability of hydrocarbon resources at a comparatively low cost. In this regard, companies and specialized institutions possessing the needed expertise and technology for increasing such resource base are hereby called upon to cooperate with our member countries, or their national oil companies, in order to achieve this ultimate all-important purpose.

On the other hand, the gas reserves of OPEC member countries are in the order of 138 billion barrels of oil equivalent, representing about 40 per cent of total world gas reserves, while the gas production from OPEC member countries is in the order of 4 million bbl/d of oil equivalent, which represents 18 per cent of the total world gas production. It is also important to note that in spite of the low production/reserve ratio for OPEC gas, a considerable amount of its production is flared. The amount of total gas production that is flared as a proportion of total gas production is about 57 per cent; the equivalent of 2.3 million bbl/d – an inexcusable waste.

It is worth mentioning that the practice of flaring gas was initiated by operating companies without considering its future utilization and its attractiveness as a clean primary energy source; in this connection no effort has been made to allocate the required investment for such utilization. Various competent companies with the required technical know-how could help a great deal in the proper and efficient utilization of such gas through cooperation with the appropriate authorities or national oil companies of our member countries. Such cooperation in increasing the hydrocarbon resource base would give our member countries greater leverage to meet the essential requirements of the industrialized world.

Due to the utter waste of flaring gas and the limited ability of OPEC member countries to consume gas domestically, a crash program of gas export facilities and reinjection projects are called for. This effort needs involvement by the consumers, due to the large investment and technology requirements. In the light of the low rentability of gas exports compared to oil (net well-head income to OPEC member countries from LNG exports is estimated at between nil and $2/bbl of oil equivalent compared to about $12–$14/bbl from oil), the best thing for our member countries would be to use gas domestically to release crude oil for exports. But in the initial stage we have to stop flaring at almost any cost.

It is evident that our present gas production is forced upon us by the gas/oil ratio of associated gas, while our future production will depend very largely on the rate of increase in gas prices, as well as our domestic consumption.

The picture is clearer when we come to analyse our potential oil reserves, production, and exports. The estimated reserves of crude oil in OPEC member countries of about 550 billion barrels could sustain a production level of 42 million barrels daily between the mid-Eighties and the end of the present century (*see Figs 4 and 5, pages 143 and 144*). Existing and planned production capacity within OPEC could sustain such a level.

However, if we do actually produce such quantities, it will result in our reserves declining dramatically at rates unacceptable to our member countries and it is doubtful whether, by then, such a sharp decline is acceptable to an orderly world economy. Needless to say, some of our member countries will suffer production declines much earlier than the general OPEC level.

However, based on the forecast demand for OPEC oil of about 33.5 and 39.3 MMb/d for 1980 and 1985 respectively[1], it is obvious that if OPEC were to satisfy this projected demand, the time span for exhausting our reserves might not be acceptable in the longer run.

Additionally, past internal OPEC consumption of oil for the period 1970–76 shows an annual rate of growth ranging from 10 to 20 per cent. If these rates of growth continue, oil consumption is bound to exceed the 3.5 MMb/d contemplated by the World Bank and may surpass the 5 MMb/d mark by 1985. Recent growth rates in product consumption in OPEC countries should not be considered an anomaly. It is natural for our countries to rely heavily on energy intensive choices and on the use of hydrocarbons as feedstock for a chemical industrial base. However, it is doubtful that rates of growth in product consumption, in the range of 15 per cent annually will fade away overnight. This, coupled with the high rate of production at present, will lead some member countries with large populations, or major industrial activities,

[1]OECD, World Energy Outlook, 1977.

such as Indonesia and Iran, to become net importers of energy in the near future, an alarming situation and one which any producer, or group of producers, would want to delay as much as possible in order to be able to save a fair share for future generations.

This analysis shows that we can all work reasonably comfortably till about 1985. We have a margin of oil export potential, as well as wealth of gas reserves, both associated and dry gas. This should, therefore, become a period of adjustment towards a more rational pricing of liquid and gaseous hydrocarbons, with the aim of achieving a smooth transfer to new patterns of energy consumption. Oil and gas supplies should, by that time, be discussed, not in terms of 'energy', but in terms of transport requirements and non-energy usage.

Turning our attention now to the revenue received by our member countries from their present oil exports, as compared to their foreign exchange needs, and in the light of their absorptive capacity, it is clear that this revenue exceeds their immediate requirements. This is best illustrated by the case of Saudi Arabia where a five million bbl/d production level could meet that country's needs, while its present production is nearly nine million bbl/day. There is no doubt that such production constitutes one of many sacrifices that OPEC member countries are making in order to honour the commitment embodied in the OPEC Solemn Declaration pertaining to the satisfaction of the essential requirements of the developed countries. The developed countries, however, should not attempt to convert this sacrifice into a weapon to combat OPEC, as is illustrated by the case of stockpiling.

The situation outlined above indicates clearly that practices and measures by both OPEC and consuming countries do not constitute the real long-term consideration of conservation, and even tends to defeat the objectives of our member countries in development and industrialization. In order, therefore, to implement a sound conservation policy, the only options at present available to OPEC are the interplay of pricing mechanism and production levels.

The various modest price adjustments since 1974, which have been aimed primarily at achieving partial compensation for the loss in the purchasing power of our revenues on account of inflation and currency variations, have been viewed by the consuming countries as a threat to the world economy and as a cause of disruption. Although it is believed that a strong upward adjustment in the prices of crude oil would provide the consumers with an added incentive to economize, encourage the use of secondary techniques, eliminate waste and intensify research in the development of alternative sources of energy, it is felt that the consuming community is not prepared to endorse such measures. However, the price and non-price related measures adopted by some major consuming countries (such as taxes and restrictions) are preventing the price mechanism from playing its full rule in the rational allocation of resources and the smooth introduction of alternatives.

In the absence of positive actions by consumers. OPEC may want unilaterally to take the necessary corrective measures by tailoring output to fit its member countries' economic needs and development programs. It goes without saying that such unilateral action would have an adverse effect on the economic growth of the world community and might result in pushing the prices up. Due to the above, and in spite of stockpiling and other aggressive measures by consumers, OPEC is refraining from following such a course.

Our Organization, while recognizing the vital role of oil and gas to the world economy, and while committing itself to meeting the essential requirements of the developed countries, believes that conservation of hydrocarbon resources is a fundamental requirement for the well-being of, and a national asset to future generations. In this regard, OPEC member countries are intensifying exploration activities with large investments to support their reserves; also, they are putting forward various plans for better utilization of associated gas, which the operating companies have been, and, in some cases still are, burning.

The industrialized oil-consuming nations, when examining their oil requirements, should not, therefore, regard the problem only as a physical obstacle, but have to consider fully the economic needs and national interests of the oil-producing countries at various levels of production. Bearing this in mind, they should not expect these producers to produce more without considering whether increased production is economically justifiable.

It is obvious from the above that forecasting future supplies of hydrocarbons is an extremely complicated task, as it is necessary to allow for a very large number of variables which cannot easily be quantified. Account must be taken of the technical uncertainties as to the volume of reserves, the real quantity that can be produced, and the efficient and profitable use of resources at present being wasted. Additionally, account must be taken of the uncertainties that exist as to the level of demand, energy policies and political decisions. All these uncertainties in short, are connected to the capability of various countries to organize their future in a rational manner, and in keeping with their basic political aims. It is very clear that both consumers and producers are interested in the possible exhaustion of oil reserves taking place at a rate, and within a time horizon, that will not jeopardize the continuity of economic activity.

To conclude, it is essential to see the role of OPEC in the future supply of oil and gas as being a case of interdependence; a road for two-way traffic. Unless sufficient note is taken of the predicament of our countries with regard to transfer of technology and development in general, OPEC should not be expected to fill an undefined role of the "residual supplier". Cooperation in our view cannot mean condemnation of our peoples to perpetual poverty and under-development. It is in the few years to come that we have to see concrete evidence of cooperation and coordination from the side of the advanced consuming countries. Since 1973, we have shown that we can supply the needs of the consumers. We have allowed the real prices of oil to fall in the face of spiralling inflation. We have taken more than our share of responsibility for the smooth functioning of the world's monetary system. We have contributed substantially, in more ways than one, towards solving the problems of our fellow developing countries. It is high time, therefore, that our other partners, the consumers, show their goodwill before the pendulum turns another full swing.

(Annex of Graphs to Chapter 3 may be found on pps. 143, 144).

# CHAPTER 4

# RELATIONS BETWEEN OIL-EXPORTING COUNTRIES AND CONSUMING COUNTRIES

*The following is the substantial text of a paper presented to the Tenth Arab Petroleum Congress organized by the League of Arab States, Tripoli, 16–22 January, 1978.*

It is not so long ago since Libya took the initiative to confront the oil companies in September 1970 and induced them to accept increases in prices and taxes. Since that date our Organization has shaken off the dust from its policy towards the oil companies and moved from its defensive position to one of positive assertion. Probably I would not be exaggerating in saying that our outstanding success on 16 October 1973, in taking pricing decisions into our own hands, in the final analysis rests on the events originally kindled by Libya in September 1970. I would like at the beginning of my talk on the relationship between exporters and consumers of petroleum to dwell a little on the starting point which led to our success in the sphere of pricing policies. The stage of positive action taken by OPEC during the period between September 1970 and October 1973 was the resultant of various factors which we should not ignore, especially today, when we are on the threshold of a stage which requires from us renewed positive action, albeit with renewed objectives which ought in turn to be pursued with new methods.

Since its birth in 1960, our Organization has been characterized by a clarity of objectives. It envisaged a central objective to which it diligently stuck for more than ten years. That objective revolved around the concept that oil prices are an integral part of the basic national interests of member countries, and that their decisions to change them are their legitimate right and one which ought not to be left in the hands of foreign oil companies alone.

Besides its clarity of vision regarding the central objective concerning prices, our Organization was also successful in setting out the proper institutional framework which enabled it to translate this objective into achievable reality. We can view the sixties as a period of continuous striving towards the formulation of the appropriate means and the creation of institutions necessary for achieving the central objective.

Through this continuous exercise and despite the differences among its members with respect to nationality, religion, and geographic location, OPEC has managed to put common interests and agreed objectives above those internal differences. OPEC also managed, gradually, and in all member countries, to promote increased national control of the oil sector in more ways than one and to subject foreign oil companies to the sovereignty of the producing state, reducing those companies from a position in which they represented independent states within our countries. It would not have been possible for OPEC to obtain its lawful rights from the companies within the structure of the then-existing framework, which allowed the companies to be on an equal footing with member countries. In this respect, our Organization succeeded in developing a process of consecutive negotiations on royalty expensing, starting in 1964, when the companies did not recognize OPEC as an international organization, and leading up to the Tehran negotiations of early 1971, when the companies themselves requested direct negotiations with OPEC as an international institution.

In talking about relations between exporters and consumers, it is no longer appropriate to reiterate generalizations. The pressing tasks ahead of us do not warrant a continuation of beating around the bush. Otherwise we shall miss a historical opportunity which, I believe, is now quite attainable – to evolve a new system of international economic relations more just and conducive to progress and prosperity in peace for our exporting countries in particular and the Third World in general. There are three immediate questions which may be put forward in this context:

(1)   What is the central objective that we want to attain?
(2)   What are the necessary means and institutions to transform the set objectives into a reality?
(3)   What is the balance of forces that makes the attainment of this objective feasible?

## 1. What is the Central Objective?

Since October 1973, OPEC has been exercising its prerogative of setting the prices of its oil exports, in the same way as other exporters of finished commodities exercise the right to set the prices of their own exports, without that prerogative being challenged by anyone.

But has OPEC really reached its ultimate goal? The answer, certainly, is "no". Since 1974 we have been confronted with spiralling price increases of materials that we import, especially those related to the requirements of economic development. We have also been confronted with a series of obstacles in all our attempts to utilize our oil revenues for the development processes. We find that the conditions for the transfer of technology to our countries are unacceptable, be it with regard to cost or quality of the technology concerned, or be it the circumstances of its usage and subsequent utilization. Our ability to export non-oil products to the industrialized countries which import our crude oil is constrained by tariffs and quotas. We find, for example, that the Arab-European dialogue has not been successful thus far in reaching any workable formula for access of petroleum and petrochemical product exports from Arab countries to European markets, at the very time when our own markets are saturated with their unhindered exports. In the Paris Conference for International Economic Cooperation (CIEC) it was almost impossible for one of the commissions to reach an acceptable wording, hinting at the possibility of increasing exports of petroleum products from the countries which are now crude oil exporters.

These indications seem to show that we have not achieved all we desire to do with our oil exports. We find ourselves today, four years after the price increase of 1973, and after exporting great quantities of our depletable oil resources, not perceptibly closer to the ultimate aim which prompted us to work for price increases in the first place. What is that ultimate aim?

The export of our finite petroleum resources is directly connected with the process of our economic transformation from countries which export a single primary commodity to countries capable of self-sustained economic growth, a perspective which must be linked in its time horizon with a period within which our oil exports will diminish or cease forever. This temporal link is exactly what is intended in the Solemn Declaration of the Conference of the Sovereigns and Heads of State of OPEC member countries held in Algiers in 1975, when it stated that:

"With regard to the depletable natural resources, as OPEC's petroleum resources are, it is essential that the transfer of technology must be commensurate in speed and volume with the rate of their depletion, which is being accelerated for the benefit and growth of the economies of the developed countries".

If our objective is economic development, with a view to establishing an economic pace capable of self-sustained growth in a time horizon which does not go beyond the depletion of our oil exports, what, then, is the particular translation of this objective?

It is to be lamented that we are still at an early stage in the formulation of common understanding on this question. It is not sufficient to say:

— we want a speedy transfer of technology with reasonable conditions
— we want reasonable access for our manufactured exports to the markets of industrialized countries
— we want the involvement of industrialized countries in the development of our human resources in operations, management, and research.

It has to be admitted, here, that there is a great deal of divergence on the content of those objectives. Not long ago, my colleague, Dr. Attiga, the Secretary General of the Organization of Arab Petroleum Exporting Countries (OAPEC), in his talk on the experience of OAPEC in the sphere of regional cooperation delivered in a Seminar held in Vienna on the role of national oil companies, stated that he "notices with regret the tendency towards individual investment decisions and the option for independent policies, a tendency which does not seem to warrant great optimism". Those comments were made in the context of assessing the capability of oil-exporting countries to seize the golden opportunity of using the oil sector to enhance their economic capabilities individually or collectively. I would like, here, to change the emphasis from pessimism into one of optimism for the future. The lessons we have learned during the past four years ought to have led us to identify minimum areas of agreement. Success cannot be achieved by forcing divergent views into a hazy frame; but by identifying a meeting point, however small, provided this convergence is absolutely clear. This was exactly the situation in formulating the original objective of our Organization when it was set up in 1960 with regard to prices. If other, secondary, objectives have been achieved since 1960, this may be attributable to the flexibility of our Organization in not forcing member countries to accept a framework where there was no unanimity. Now that we are at a new cross-roads, we must formulate our objectives with modesty, but at the same time, with absolute clarity. Otherwise we shall not be successful.

I believe I am not saying anything new when I point out that the advanced industrialized countries that import oil have already set their own objectives very clearly. They want from us the security and continuity of oil exports at prices acceptable to them. In essence, they want to determine the quantities that we ought to produce and the prices at which those quantities are sold to them. When we proved our capability of setting prices and production levels they requested that we should involve them in this prerogative of ours. Against this, what are they offering us? Nothing much so far. What means and institutions have they set up to achieve their own objectives?

They have set up – besides their many and inter-linking organizations like the EEC and OECD – the International Energy Agency and have given it various powers involving all aspects of the oil industry. Within the framework of this agency they have signed many detailed agreements. On the top of all this, they have substantial leverage over oil companies in varying degrees.

In the face of this clarity of objectives and precise evaluation of means, we ought to terminate the stage of general discussion about our requirements from the industrialized countries. The most appropriate way to put an end to the debate is, not to force divergent interests together, but to search for their most common denominator, and, surely, we have many economic features in common; then we ought to look for the best formula for cooperation which allows our objective to grow and expand in a natural manner. It is from this standpoint that I find my justification for optimism in the future, because I believe it will not be impossible for us to find common interests between us, once we have the political will to do this.

2.   *What are the necessary means and institutions?*

The developing countries are carrying on bilateral and multilateral discussion with the developed industrialized countries, such as the Arab-European dialogue. It might have been possible for the North-South Dialogue in Paris to become a practical forum for negotiations because of its flexible nature, but unfortunately the industrialized countries decided to scuttle that dialogue. Probably the reason for the unyielding attitude in Paris was due to their anxiety about the strength of the alliance between the Third World countries in a confrontation the like of which they had never before witnessed. The developing countries have also sought to bring about a forum for direct dialogue within the framework of the United Nations. The establishment of the United Nations Conference on Trade and Development (UNCTAD) was a source of a great deal of optimism in 1964. But the past thirteen years and, particularly, the failure of UNCTAD to achieve the integrated programme for commodities and its special fund, make us tend to believe that existing fora within the United Nations are not sufficient for speedy programmes to save the countries of the Third World from backwardness. I do not want here to belittle the basic and necessary role of the existing institutions of the United Nations, but I do want to point out the need for developing nations, and among them, our oil-exporting countries, for new fora which are more effective and capable of achieving speedy fulfilment.

Some of our member countries belong to various regional organizations, such as the Association of South East Asian Nations (ASEAN), the Organization of Latin American Countries for Energy (OLADE), the Organization of African Countries (OAC), the League of Arab Nations, OAPEC, etc., all of them organizations which have interests in economic matters and which look after the interests of their member countries including those concerning development. Nevertheless, they also allow for the possibility of evolving a new formula for solving the particular problems of oil-exporting countries.

The opportunity may arise in the near future for setting up an appropriate framework for the dialogue with the industrialized countries, if the conditions for a common vision and objective evolve. It is also possible that various frameworks on a regional basis may be set up with some central coordination. However, the requisite for the success of any possible framework is clarity of vision and the appropriate choice of negotiating platforms. It is also necessary to point out that the various means and institutions for negotiations with the       industrialized countries will not achieve much success unless they are linked in one way or another with the capability for counter pressure. We must not forget that politics is the science of the possible and that economics is the central theme in politics. Therefore, the industrialized countries would only give way to the extent that the countervailing pressure of the other side would necessitate, and in accordance with what the other side might offer in response. The developing countries, despite the greatness of their number, and the weight of their population and area, could not force the industrialized countries to sit at the negotiating table until a group of developing countries, i.e. the oil-exporting countries, had an effective means of pressure on the industrialized countries. It is here that the Paris Conference, despite its failure, has its greatest significance in the sense that it showed our weaknesses and strengths. This reality has to be kept in mind in future encounters.

One of the practical fora for cooperation and exchange of views exists in the evolving relationships among national oil companies of OPEC member countries. These companies play a very significant role in investment and development operations in the oil sector in its narrow sense, and outside that sector, too. It is hoped that cooperation between national oil companies of OPEC member countries will evolve to the extent that they constitute an important vehicle for the transfer of technology with acceptable conditions. They are capable at least of coordinating their activities in the direction of reducing competition among themselves and formulating common bases for the supply of materials and the execution of projects within their spheres. The importance of this level of cooperation is not diminished by the fact that it represents only a limited number of developing countries, i.e. member countries of our Organization. If we succeed in manufacturing petroleum products in the direction of refining petrochemicals and fertilizers, this may provide an impetus for other commodity or regional groupings to formulate frameworks and institutions appropriate to their own situation.


3.   *Balance of Power*

We have pointed out that the industrialized countries want from us secure supplies of crude oil at prices acceptable to them. As time passes by, it seems clear that those countries cannot do away with their needs for our oil exports, even in the long run.

If they are successful in all their plans to increase coal production and the construction of nuclear power stations, which is a matter of some doubt, they will continue to import quantities of oil in excess of their present imports for a period which might last till the end of the present century. Despite all their efforts to restrict energy consumption, and with the most optimistic forecasts, they are striving to maintain the present level of their oil imports from OPEC countries. Briefly, and without going into the details of well-known divergencies on future energy demand, it may be safely said that demand for OPEC oil will not diminish in absolute terms, except the natural decline of those exports under the pressure of depletion in the long run. Despite this significant negotiating advantage, we ought not to exaggerate our ability to enforce our own conditions on the industrialized countries, since they are in a better position to coordinate their policies, and their dynamic economies are capable of adjustment to an extent which we cannot ignore. Therefore, when we evaluate our experience of price adjustments since 1973, we ought to recognize the fact that the economies of the industrialized countries have been successful, to a significant degree, in circumventing our price decisions and depriving them to some extent of their content. We find ourselves today, after four years, in an inflationary situation which encroaches heavily on our purchasing power and we find that our financial surpluses are at the mercy of industrialized countries. I had previously pointed out in a lecture delivered at Harvard University that we control only part of the real price decisions; probably we play a less significant role compared to that practised in fact by the industrialized countries. Those very countries ultimately determine the final price of petroleum products, and their share of the cost of petroleum to the final consumer is greater than that of the exporting countries. Furthermore, they control the decisions of pricing the commodities they export to us, on which we spend our oil revenues, etc.

Within the balance indicated above, we must realize that we have to utilize our negotiating position to the maximum and achieve our aims in a creative manner. Probably it is also necessary to bring into the balance two other elements which can help us to gain the upper hand.

(1)  Our commercial weight as markets for the exports of the industrialized countries.

(2)  The special position enjoyed by the financial surpluses of some oil exporting countries.

Probably the most important element in the overall situation relates to the science of timing the seizure of the historical opportunity available to us to exert pressure on the industrialized countries in order to achieve the necessary conditions for our transformation from backwardness to progress.

When we go into the details of statistics we find that the production of oil by OPEC countries may not exceed 40–45 MMb/d if this production level is to be sustained for some period of time. Within this level, some member countries are producing at levels near their maximum capacity. This means that the necessary increases must come from certain countries only, and that continuous pressures on the production of those countries will ultimately lead to forcing their production levels down. Some experts estimate that OPEC production will tend to decline, in absolute terms, around the end of the present century, while the production of some OPEC member countries will decline much earlier. If we take into consideration the increasing domestic consumption of petroleum products in our countries, we find that what remains for export is indeed very little for some member countries. If we combine the internal consumption of our member countries, and if that consumption continues at rates approaching 15 per cent annually, or even if those rates decline later to 13 per cent or 10 per cent annually, the combined internal consumption of OPEC countries may reach about 20 MMb/d by the end of the century. This would mean that our exports would tend towards natural decline in the last decade of the present century.

On the other hand oil-importing countries, and especially the members of the International Energy Agency (IEA), are vigorously striving to reduce their dependence on their imports from OPEC countries. They are following various means of weakening the bargaining position of OPEC. That group of nations is building up stockpiles of oil which will cost them about $100 billion, an amount equal to the value of all their imports from OPEC for a whole year. All this on top of the fact that the degree of the industrialized countries' dependence on oil imports to meet their energy requirements will be reduced greatly, since they are putting a great deal of effort into developing alternatives to imported oil.

In a nutshell, and with regard to the balance of power, we have to take note of the following factors:

(1) The negotiating power of some OPEC member countries vis-à-vis the consuming industrialized countries will tend to diminish gradually to the extent that the oil exports of those countries will face a natural decline or be required to meet increasing domestic consumption.

(2) The negotiating power of OPEC as a whole will tend to diminish gradually if we measure it by the degree of dependence of the other party on oil imports from OPEC.

(3) The strategy of industrialized countries centres on reducing the reliance on OPEC in various ways.

In our estimation, therefore, the industrialized countries will continue to depend on OPEC oil in the short and medium terms. Their need for OPEC oil in the long run may continue, but OPEC exports themselves might diminish by then. Furthermore, the dependence of industrialized countries on OPEC oil specifically, may decline. Therefore, we should not feel relaxed only to be jolted some day by the creative ability of industrialized countries to rid themselves of their reliance on us; then the words of the glorious Koran would come true in our case: "We have not done them an injustice, but they have done an injustice unto themselves".

Probably, in the best of circumstances, even if we commence our development efforts today, such development would need more time than the remaining years of the present century. To construct an infrastructure for our economies including ports, highways, communications networks, universities, etc., we would need a decade or more. Even if we begin to set up factories side by side with the establishment of the necessary infrastructure, this would not obviate the need for the following stage of coordinating the various economic activities, balancing them, and filling the time gaps arising in their establishment. Furthermore, if we are to have economies capable of self-sustained growth, we need to prepare the human resources and the necessary administrative capabilities for an advanced society. Here, probably, lies one of the biggest deficiencies, one which needs more time than any of the links in economic development.

A French expert in the context of technological transfer indicated that the College of Technology in Tehran took ten years to mature and that possibly another ten to twenty years are needed to establish a proper research centre. The opinion of such experts may or may not be precise, but the moral is that unless we start now, we may condemn our future generations to poverty, and history will judge us to have failed miserably.

### 4.  Subjects Under Consideration

There are a large number of subjects that deserve to be put forward in negotiations between oil-exporting countries and industrialized importing countries. The identification of such topics requires a clear, practical plan of action capable of execution within a limited period of time. As far as oil-exporting countries themselves are concerned, and within the framework of our Organization itself, we are facing a new situation and new challenges which are different from those we faced during the sixties. It is necessary, therefore, to reassess our traditional strategy and methods. I will try here for the sake of greater clarity of objectives to summarize briefly the main topics raised in this context, with a preliminary evaluation of each of them.

## 1.  Prices

There are several studies and evaluations indicating the proper level of prices and the correct basis for determining them, whether in relationship to prices of petroleum products in the industrialized countries themselves, or in comparison with the various alternatives to the uses of such products, or by relating them to the scarcity of liquid and gaseous hydrocarbons in nature. However, we are not here discussing the just levels of crude oil prices and the means of reaching them. What is important in the final analysis is the real value of what we get in lieu of spending our oil revenues and, especially, what we get in relationship to development. The most important element of the price, therefore, is not the number of dollars, but the real development which is attained by exporting countries against the quantities exported by them and the reserves depleted thereby.

On the other hand, our oil revenues are affected by the terms of trade, i.e. the relationship between movements in crude oil prices and movements in the prices of commodities imported by OPEC member countries. We have noted that increases in the prices of commodities exported to OPEC countries are about three times as high as the increases in the prices of exports of the industrialized countries to each other. An end has to be put to this vicious circle in relations with the industrialized countries. But in order to achieve reasonable progress in this area, we must first study our own import organs in an objective manner. We also need to examine bottlenecks in our facilities such as ports, highways, storage, as well as the organs responsible for the execution of projects. It is only by assessing the pros and cons that we can find the proper solutions to avoid internal competition among ourselves as a minimum necessary level for improving our negotiating position in the area of imports. From there we can go further to discover the appropriate means to coordinate our trading operations in order to face the coordination which already exists between the companies that control exports to us. When we reach such a situation of utilizing the spheres of action available to us, then the problem in its real perspective can be put on the negotiating table with the industrialized countries concerned.

## 2.  Transfer of Technology

From among the central issues of contention with the oil-importing industrial countries, the questions of development and, especially, the transfer of technology, are the most conspicuous. However, it is about time to spell out precisely what is meant by "transfer of technology". Technology cannot be given as a gift, nor can it be bought as a piece of machinery. Probably the subject of technology acquisition depends more on those who receive it than on those who give it. We have to prepare the proper ground to foster technology in our countries, starting with its simple basis, in which we do not have the need for recourse to negotiations.

Expansion of education, the training of manpower in basic skills and the introduction of relatively simple modern techniques in our production activities do not require us to request them from a foreign party. Furthermore, the establishment of basic industries which are suited in their technological complexity to our limited abilities does not need a great deal of effort. We can, therefore, limit the question of technological transfer to the very complex production techniques which are still monopolized by certain companies. If we have to resort to such technological monopolies for specific purposes, like certain aspects of the petrochemical industries, we should be prepared to the fullest extent with regard to skills related to the subject which are not monopolized, in order to limit the area of negotiation on the one hand and to prepare the necessary climate for the absorption of the technology concerned on the other. If we cannot conduct this process of absorption we will always be tied down to foreign technology, even it it is available to us on reasonable conditions as we request them.

A subject related to the transfer of technology with a wider scope is the speedy development of OPEC countries themselves. For the development, we need first to evolve economic plans in each country, which take into consideration the requirements and capabilities of that country. Then, those different development plans could be coordinated and areas of complementarity between them identified. This process would help us to get fair conditions from the industrial countries to execute the necessary develop-ment projects and to avoid overlaps between these projects on the one hand, and to achieve appropriate prices and specifications on the other.

### 3.   Access of our Non-oil Exports to Markets

The anxiety about the ability to market petrochemicals, petroleum products, and other manufactured goods might be considered as the biggest impediment facing our development objectives and it represents the most complex difference between us and the industrialized countries. The experience of the Paris Conference, the Arab-European Dialogue, and the recent Seminar held by our Organization in Vienna, on the role of our national oil companies, all seem to indicate that the industrialized countries are not prepared to make basic concessions in this area. Their well-known claim is that there is excess capacity in refining and petrochemical industries in the world. Despite the existence of some reality in this claim, it is to be noted that the industrial processes which take place in the Third World, or even in Japan, would never have manifested themselves if they had been carried out on the basis of acceptance by the industrialized countries. The only way for us to proceed seems to be by going ahead, since the industrialized countries will never refrain from bringing about increments in refinery and petrochemical capacities unless we ourselves make them first.

At that point the subject for negotiations will not be whether to industrialize or not, but it will center on the proper division of markets between our products and theirs. This is also applicable to a large number of our anticipated manufactured exports. In this respect we should not ignore the importance of our imports from the industrialized countries, which may play a big role in getting some countries to accept our manufactured exports in exchange.

### 4.   Energy Matters

Some advanced industrialized countries try to get OPEC into a direct dialogue on energy matters. This would mean drawing OPEC into negotiations on its incontestable, legitimate right to determine the prices of its exports and the quantities of such exports. The industrialized countries will never draw oil-producing countries into negotiations on these subjects in any framework. If the industrialized countries are ready to involve us in their discussions on the pricing of their various exports, as well as the destinations and quantities of such exports, and allow us to participate in their own plans to produce their indigenous alternatives to oil imports, only then would there be some complementarity in such negotiations. Otherwise there are no reasons to induce us to participate in unequal negotiations.

The industrialized countries are seeking to evolve their own energy policies with one common objective, namely the reduction of their dependence on OPEC oil and gas. They are at the moment conducting frantic efforts to bring about all means available to them to reduce the importance of oil exports, even if this causes some imbalances in supply and demand. How can we, in this context, talk about direct negotiations on energy? At a time when OPEC is doing all it can to take the interests of the industrialized countries into consideration, we are witnessing that the latter are ignoring the interests of OPEC.

It might be appropriate to raise here a possibility of OPEC participating directly in the field of alternative sources of energy. In this way we would be in continuous, direct contact with the market possibilities and their effect in the long run on the future of crude oil. This would also help us to consider, as of now, the interest of member countries that may resort to importing energy alternatives in the not-too-distant future. Furthermore, OPEC countries have great capabilities in the field of solar energy. There does not seem to be any justification, therefore, for the exclusion of OPEC from the areas of research and development in alternative sources of energy.

## 5.  Financial Surpluses

Some of our member countries have financial surpluses in their current accounts resulting from the export of quantities of oil in excess of their immediate needs of expenditure. It is worth noting that these surpluses are not finding sufficient guarantees or remunerative investment opportunities in the oil-importing countries. Despite the fact that these surpluses arose as a result of OPEC's efforts to meet the requirements of consumers for crude oil, it may be appropriate that the member countries concerned with these surpluses should coordinate their efforts to get appropriate conditions for their investments. Changes in exchange rates and monetary inflation affect directly the real value of these surpluses. Furthermore, the investment opportunities existing for such surpluses are not meeting preferential treatment from the industrialized oil-importing countries. It should not be difficult to find a satisfactory formula for coordinating the efforts of the countries concerned to arrive at minimum conditions acceptable to them which could be put forward collectively to the industrialized countries in this respect.

Following this quick identification of some subjects of contention between the industrialized countries and the developing countries in general, and between the former and OPEC countries in particular, I would like to stress the central issue in international economic relations.

We have witnessed the collapse of the Paris Conference as a result of the unfortunate insistence of the industrialized countries on a position which reflects a selfish, narrow point of view. More recently, we saw the collapse of the efforts of UNCTAD to establish the Special Fund for Raw Materials. There seems to be a void and a temporary pause in the dialogue between the industrialized countries and the developing countries. Unless the world finds a solution to the big gap between the rich and the poor, this situation may lead to continuous tension in the world situation which may deteriorate to an explosive extent that would be harmful to human civilization. The experience of our world since the industrial revolution, and the experience of the industrialized countries themselves, indicate that prosperity increases and expands to the extent that increasing numbers of countries and regions participate in it. I do not believe that the industrial programmes of the Third World countries will lead to an impoverishment of anyone. On the contrary, international exchange would necessarily increase world prosperity and open up wide avenues for material welfare.

Finally, I would like to say that the survey I have presented on the relations between oil-exporting countries and the importing industrial countries indicates that there is a great deal of vagueness in what we reiterate with respect to these relations. It is time for us to re-evaluate our standpoints on these subjects, to precisely identify anew, and to devise the appropriate institutions for achieving those objectives. I reiterate my optimism that it will not be impossible for us to find a minimum common denominator. This, in turn, does not exclude the efforts of any regional or bilateral grouping to achieve objectives beyond the minimum general level, provided that all our efforts, whatever the framework and institution, all pour into one stream.

# CHAPTER 5

## PROBLEMS AND PROSPECTS OF STATE PETROLEUM ENTERPRISES IN OPEC COUNTRIES

*The following is the substantial text of an address delivered at the United Nations Inter-Regional Symposium on State Petroleum Enterprises in Developing Countries, Vienna, 7–16 March 1978.*

### I

It gives me great pleasure to address this important Symposium. As you all know, last October we held a Seminar at our Headquarters here in Vienna on "The Present and Future Role of National Oil Companies". This Seminar concentrated mostly on problems and prospects of the National Oil Companies in OPEC member countries.

Although the topics raised during the OPEC Seminar might not be absolutely identical with those discussed in this present forum, the area of common interest is very wide indeed. OPEC countries face the same problems which other developing countries encounter, especially with regard to development needs.

Our group of countries relies very heavily on the production and export of crude oil and gas, and therefore, the state petroleum enterprises in OPEC occupy a special place in the economics of our member countries. However, we realize that in the wider group of developing countries, there are state petroleum enterprises which have a longer history and, in some cases, more comprehensive experience than our National Oil Companies. We have learned, and are still learning, a great deal from the experience of enterprises such as the Mexican National Oil Company (Pemex), the Indian Oil Corporation, the Indian Oil and Natural Gas

Commission, Petrobras of Brazil, and others. Indeed, some of our National Oil Companies are only in their formative years.

Only since the beginning of the seventies have the majority of OPEC National Oil Companies, (NOCs) been closely involved in the various aspects of the oil industry. However, despite their relatively short experience, these Companies have been playing an increasingly important role in the international oil industry and, in many cases, have taken the leading role in managing the business of producing and exporting oil in their respective countries.

Looking at the historical growth of our NOC's, we see that some are older than OPEC itself, while others came into existence after the birth of the Organization in order to cater for the special circumstances of an individual member country, i.e. Algeria after independence, Iraq after the passing of a law in 1961, etc. However, their real birth and growth came after the 1970s with the demand for participation on the part of the existing operating companies and, as we see it, the task has been well accomplished. However, the success achieved has not been without its problems and difficulties; neither complete transfer of legal ownership, nor the future growth and development, could ever go as smoothly as is planned on paper.

In looking at the problems which were encountered, we have to go back to the early days of the oil industry when the concession agreements concluded between the international oil Majors and host countries in effect gave the companies carte blanche to carry out their operations anywhere in the country's territory. This effectively elbowed any nascent National Oil Companies out of all but the most unpromising fields. The advent of the so-called Independent Oil Companies improved matters slightly, but the pattern remained largely unchanged.

Indeed, although many concession agreements accorded a theoretical right for each state to receive in kind the royalty crude (about 12½ per cent of the oil produced in most countries), other conditions stipulated that it should be received at the posted price. Many countries, however, found that they would stand to lose in so-called "profit sharing" arrangements if they did not receive their royalty crude in cash. Due to this arrangement, national oil firms had no access to any crude oil with which to compete in a controlled market in so-called "arm's length deals". Indeed, it can be said that marketing experiences in many cases are to be learned simply from reading about them.

The legal takeover of the oil industry from the international Majors, which we have witnessed in recent years in OPEC countries, has entailed increased responsibilities, and in some cases, problems too. We are now standing at new crossroads in our history as an organization and as a group of countries. We are facing a fresh challenge to formulate a new vision which takes us beyond the legal control of exploration, production, and export of crude oil.

As I said last October during the OPEC Seminar, our new vision could be based on the following objectives:

1) The efficient management of the oil industry by nationals of OPEC member countries at all levels;
2) the development of an indigenous technological base, backed by domestic research institutions, capable of contributing increasingly to the needs of the oil sector in member countries;
3) the speedy transformation of the role of OPEC member countries from that of raw material exporters to manufacturers by carrying out certain downstream operations, especially with regard to refining and petrochemicals. In this way, the national oil-industries should become the central pivot in the process of industrialization.

Obviously, this formulation means that the state petroleum enterprises in OPEC member countries ought to play an increasingly important role not only in the economies of their own countries, but also in the formulation of the policies and objectives of our Organization.

## II

If our NOCs are to meet this considerable challenge, there are a number of problems which they will have to resolve. In many ways, these differ very little from those facing other state petroleum enterprises in the rest of the Third World. I will, therefore, try to identify briefly some of those problems, in the hope that this exchange of ideas may help us to go some way towards resolving them. Needless to say, the member countries of OPEC differ in terms of the nature of their economies and also of the degrees of direct involvement in their indigenous oil industries.

*Manpower*

Some National Oil Companies control the whole of the oil sector, others depend, to varying degrees, on the cooperation of foreign companies. However, the speedy rise to eminence of OPEC NOCs has created serious personnel and manpower problems in all our member countries.

There is a lack of suitable trained, experienced and specialized personnel at all stages necessary to cover the administrative and technical operations of the oil industry. The root of this administrative and manpower problem lies in the former policies and attitudes of foreign operating companies towards personnel recruitment coupled with the limited national human resources in certain cases. To face this serious shortage, OPEC NOCs have been seeking various interim remedies to bridge the gap.

Some member countries are relying heavily on foreign personnel and operating companies. In the meantime, intensive training and development programmes are being carried out in all skills needed for the oil industry. This task is extremely difficult and complex, since it involves a country's whole educational system and not only the organization of a specialized institute. Most of our member countries endeavour to send some of their nationals abroad for higher education courses and management programmes at foreign universities and institutions, so that at a later stage they can be of benefit to the industry either directly, or indirectly, via the educational system.

However, some of our member countries are already facing two problems which might hamper their ambitions in this direction. The first and lesser problem is that of the brain drain to more advanced countries which afflicts our countries in the same way it does all developing countries. The second and wider problem is related to the ambitious economic development plans of our member countries which compete with the oil industry for qualified manpower. Despite the special status of the oil industry in OPEC countries, it cannot stand in the way of general economic progress by monopolizing the limited managerial and manpower resources available.

*Operational Problems*

The transition from foreign control of production in OPEC countries to increasing national control has been fairly smooth, even in cases where complete nationalization has taken place. In all OPEC countries no serious disruptions have been observed in any of the upstream phases of the oil industry. But this does not mean that we are happy with the present state of exploration and production in our countries. There is a slight slowing-down in exploration activities in OPEC member countries, despite the fact that they still hold better prospects of new discoveries than many other areas. We are particularly unhappy with the old practice of the foreign oil companies of gas flaring, indeed our National Oil Companies need to make a tremendous effort to put an end to all forms of flaring of associated gas. In this respect and to varying degrees they need involvement with and cooperation from the consuming countries, both with respect to technology and the appropriate pricing of gas.

With the increasing demand on OPEC oil resources, many of our member countries are resorting to secondary recovery techniques and other enhanced recovery methods to maintain production levels and to maximize ultimate recovery. Some of our member countries are allocating a great deal of their oil revenues to investments in enhanced recovery, and here too our National Oil Companies are in need of cooperation from the consumer side.

*Refining and Petrochemicals*

The economies of developing countries generally suffer from dependence on the export of a few, or one single natural resource. Those countries which depend on the export of a depletable natural resource are in a particularly awkward position. They must achieve their emergence from perpetual poverty before the depletion of their natural resources, or even before total depletion is reached. Ideally, these natural resources should become raw materials for a sound industrial economy. In this context, it is only natural that OPEC countries should look to their refining and petrochemical industries as engines of change, playing a central role in their industrialization policies.

Unfortunately, the stepping-up of refining and petrochemical capacities in OPEC member countries is facing tremendous resistance from the developed countries. This resistance, if it persists, will lead us to cast serious doubts on the willingness of the rich nations of the world to cooperate in bridging the great divide between rich and poor.

## Internal Energy Needs

State enterprises responsible for internal energy requirements in OPEC member countries are surprisingly facing some unexpected difficulties in their own house. For instance, due to rises in energy consumption, there are imbalances in energy supply and demand in some member countries. The short-term imbalances are probably a healthy sign of change, although they cause serious inconveniences. However, in the longer term, difficulties could arise as a result of pressures from the domestic needs for energy and non-energy uses of oil and gas. The energy-intensive industries now being developed in our member countries as well as the petrochemical complexes envisaged for the future, might face serious shortages in the long run, if depletion of petroleum resources for export is not checked in member countries with limited reserves.

## III

Having briefly examined some of the problems now being faced by OPEC NOCs, I would like to say a few words about the role which National Oil Companies in OPEC member countries, and indeed in all developing countries, may fulfill.

The setting-up of a National Oil Company has become very fashionable these days and in some quarters doubt has been cast on the need for NOCs, comparing them with superfluous status symbols like national airlines. However, in the case of OPEC countries, National Oil Companies were created to meet a greatly felt need. During the fifties and most of the sixties, the transnational oil companies dominated the oil scene in our countries. They were in most cases not subject to domestic clauses and exercised the power of a state within our countries. The oil sector was completely isolated from the rest of the economy and basic decisions were taken in the international centres of these companies. The only possibility for our countries to have a greater say in the ultilization of our basic source of livelihood was to create national institutions which could ultimately replace the foreign oil companies in every way possible. These new institutions had to start from scratch in a hostile environment and addressed themselves to the incredible task of replacing the biggest and most complex industrial giants of modern times.

In countries like Venezuela, Iran, Iraq and Algeria, some of the basic skills were already available to create national oil enterprises which could go straight away into the oil business. In other countries, the enterprises had to feel their way at a slower pace. The international Majors treated these new arrivals with a degree of aloofness and hostility. However, several independent companies and national companies of consuming countries were prepared to deal directly with the OPEC National Oil Companies. These joint ventures and contracts of the sixties and early seventies helped a great deal to strengthen the hand of OPEC NOCs and to give them the necessary confidence for the future.

Now, in the latter part of the seventies, OPEC NOCs are oil giants in their own right. We are entering a new phase where the National Oil Companies play a predominant role not only in their respective domestic oil sectors, but also in the national economy as a whole. There is no doubt now that OPEC NOCs will become the backbone of the upstream part of the oil industry, as indeed they are already in some countries. Due to their very nature, they have tended to serve the development efforts of their countries and, in some cases, have succeeded in integrating the oil sector into their national economies. In other cases, they are involved in development efforts outside the oil sector, sharing not only in the processing and industrialization of crude oil and natural gas, but also contributing to general industrialization efforts in other sectors, due to their experience and successful institutional structure.

The new phase in oil industry development, to which I referred earlier, is characterized by the perceived need for international interdependence in the energy field as a whole and, indeed the industrialization process in general. The era of direct conflict with the international Majors is now well behind us and a new relationship is emerging between National Oil Companies in OPEC countries and foreign countries. The basic feature of this new relationship is greater cooperation in several fields, especially in exploration, development, and international trade of crude oil and gas. This does not mean, of course, that there are no differences of opinion and areas of conflict with transnational corporations, especially with regard to downstream operations, such as refining and petrochemical development.

It is also worth noting that the cooperation which started in the sixties between OPEC countries and National Oil Companies of the consuming countries is still growing in a period of increased governmental involvement in energy policies. State petroleum enterprises of the consuming countries are serving as useful instruments in direct deals with OPEC NOCs. This is true only as far as state petroleum enterprises of industrialized countries are concerned, but also for enterprises in the developing countries.

It is important in a forum such as this to consider too the relations between OPEC NOCs and state enterprises in energy deficient developing countries. It must be said that this area deserves a great deal of consideration and some deep re-thinking is required on both sides to evolve the best form of relationship which serves the mutual interests of oil-exporting countries and oil-importing developing countries.

Although they are not responsible for the main diseases of the modern international economy, i.e. inflation and unstable monetary systems, OPEC countries have contributed substantially to alleviating the balance of payments problems of other developing countries, despite the fact that some are in the capital market themselves to support their own development programmes. But our relationships as developing countries should go much deeper than mutual assistance. In several cases, OPEC NOCs have contributed to the development of the energy sectors in other developing countries, such as financing refinery constructions, ensuring flexible oil supplies, exploration agreements, etc.

In the cases where state petroleum enterprises in the oil-importing developing countries are well established, such as in India, Turkey and Brazil, cooperation is taking on very concrete and fast evolving forms which have already led to some satisfactory results. However, the resources of OPEC NOCs are very limited and, as we noted earlier, their domestic problems are many.

There is, in my view, no readily available formula to resolve this problem, but it is hoped that state petroleum entities of the developing consuming countries can improve their capabilities. I am sure that they will find OPEC NOCs ready to cooperate to their mutual satisfaction.

Before concluding my remarks, I would like to say that OPEC NOCs are still feeling their way to maturity and in some cases there is still a long way to go. However, they are embarking on increased mutual cooperation in the various fields of their activities. This cooperation is in its very early stages, but it is hoped that it will evolve speedily and point the way to the type of cooperation needed among state petroleum enterprises in developing countries, whether they are exporting, self-sufficient, or importing petroleum.

There is no reason why this cooperation should not go on to contribute to the greater economic independence of the Third World. The traditional relationship of unequal exchange with the industrialized world should give way to the new pattern of mutual cooperation now evolving among the developing countries. This cooperation and solidarity should be the corner-stone of our collective demand to bring about a New International Economic Order.

CHAPTER 6

# AN OUTLOOK FOR THE OIL INDUSTRY
# AND THE ROLE OF OPEC

*The following is the substantial text of an address delivered at the First International Conference of Petroleum Investment Analysis, London, 31 May, 1978.*

In the past few years interest in the subject of long-term supply and demand of energy in general, and of oil in particular, has been growing. A number of institutions are devoting a great deal of time and effort to forecasting future world energy requirements. With the wealth of literature that has been coming out, especially during the last two years, it is very difficult to keep up to date with the latest forecasts, let alone to make critical evaluations of all of them. One need not add to this proliferation by presenting my own forecast of energy supply and demand. The best course of action is probably to look into the existing forecasts and try to identify what the most likely outcomes will be in the light of the assumptions on which these are based.

For us, the most important aspect of energy demand is what is termed as "production from OPEC". Various forecasts put this demand between 33 to 35 MMb/d by 1980, and 36 to 52 MMb/d by 1985. The latest forecast by Exxon, which came out in April this year, estimated demand for OPEC oil at 35 MMb/d by 1980, 40 MMb/d by 1985, and 44 MMb/d by 1990.

If we look into the supply side, we find that such levels of demand could be met, especially if potential OPEC reserves materialise in the presently anticipated magnitude. OPEC member countries now hold 70 per cent of proven recoverable world reserves. It is also worth noting that OPEC member countries will contribute substantially towards the additional recoverable oil and will account for nearly half of the world's ultimate recoverable reserves. They also contain nearly 40 per cent of proven gas reserves, and it is further estimated that they contain about 30 per cent of all probable further additions to world gas reserves.

Taking into account the share of OPEC in the estimated ultimate recoverable reserves and bearing in mind the ease and lower cost of bringing in the new oil, be it from new fields or by using enchanced recovery techniques, OPEC's capability of meeting the estimated oil demand will improve, thus fending off the anticipated oil crisis. The upper limit on OPEC's potential production should be such as to enable OPEC to accommodate anticipated demand for its oil up to the end of the present century. I have previously estimated this upper limit of OPEC potential production to lie in the range of 40 to 45 MMb/d.

However, there are a number of constraints which might act alongside the geological limitations referred to above, which limit OPEC's potential exports in various ways. One of the important elements that we have to take into account is related to the approach made by OPEC member countries regarding the issue of conservation on the production side, from the point of view of optimum reservoir management, and even more so from the point of view of intergenerational equity. Some member countries might feel that their interest is best served by extending the lifespan of reserves beyond this century in order to safeguard the interests of future generations, which automatically means lowering the ceiling of production potential. Another factor, influencing exports rather than production, arises from the rate of domestic consumption of petroleum products for energy purposes, as well as their use as raw material input, particularly in the chemical industries. From a general examination of various forecasts, it seems that there is a tendency by the consuming countries to underestimate future domestic consumption of petroleum by OPEC countries. Due to the use of intensive energy industrial projects and reliance on petroleum products as a raw material input in the industrialization process, our member countries are witnessing very high growth rates for domestic consumption of petroleum products.

Preliminary estimates indicate that by the turn of the century OPEC countries might be consuming substantially higher than those estimates suggest. If such high level of domestic consumption is reached, the export potential of OPEC, as a group, would be very severely constrained. It remains to be said that OPEC has a very significant potential for gas production and exports. The contribution of associated and independent gas towards meeting world demand for energy will, to a great extent, depend on the evolution of gas prices, and the technological and marketing assistance rendered to us by the industrialized countries. At present OPEC accounts for 20 per cent of world gas production, but unfortunately 35–40 per cent of it is flared. However, gas will contribute substantially to the future internal requirements of OPEC countries in any case, and thereby indirectly liberate additional crude oil to be exported.

It is not easy to conclude a clear supply-demand relationship from this general picture, but it is obvious that the possibility exists for arriving at a harmonious evolution of supply and demand, provided the circumstances for such harmony are put in motion at an early stage. The assumption that OPEC will fill the gap assigned for the residual supplier under any circumstances, and in all eventualities, is a gross over-simplification. We have to examine a very wide range of possibilities and eventualities that might take place in reality, within the boundaries we have touched upon earlier.

It would appear that the consuming countries seem to assume that OPEC should produce in accordance with their whim and wish, production fluctuating substantially to satisfy the consumers' policies, with no due consideration being given to the budgetary needs of OPEC member countries. The general attitude is to consider OPEC as a sort of unconstrained warehouse which does not even charge the users storage costs. They also assume that they could have access to the contents of this warehouse regardless of the reactions or needs of the 320 million people in OPEC member countries. The idea that the people of those countries have no will, or no national interests to look after, is obviously not only callous but could invalidate any calculations.

Now, when looking into the present behaviour and attitudes of the consuming countries, a catalogue of measures and policies can be found which can only result in great tensions in the future. Also to be noted is the continuation of concerted action explicitly designed against OPEC interests, or at least aimed at reducing OPEC's role. Strategic stockpiling is being pursued at exorbitant costs for this purpose, while conservation measures are very often distorted from their basic purpose of improving the efficiency of energy use and are implemented in a manner which can only hamper the drive towards its efficient and rational utilization. We also note that tariffs and taxes on crude oil and products take forms which can only be interpreted as a circumvention of rational pricing of this scarce and exhaustible raw material and as barriers against potential OPEC exports. It has often been cited that, while the final consumer pays up to $40 per barrel, the price of the raw material is declining in real terms. In this era of concern over energy conservation and potential energy shortages, the non-remunerative pricing of gas is the direct cause of so much flaring of associated gas. The catalogue of selfish and one-sided policies is very long. One of its aspects presently influencing the oil market and greatly hampering energy planning, is the uncoordinated production of oil from the North Sea and Alaska, which has resulted in a so-called oversupply of oil.

These new supplies, coming in after a period of economic stagnation, are intended to result in a continued reduction in the real price of oil. On the other hand, when a shortage develops in the market, OPEC is expected to avoid any price adjustments or increases. However, if we are falling back into another prolonged period of declining real oil prices similar to those of the 1960's, then a reaction such as that which occurred in the early seventies could very well repeat itself if we are not careful.

In the long run, if the supply-demand balance is to be met, intensive exploration for oil and gas should concentrate on developing OPEC's potential rather than the present trend of disproportionate and excessively costly exploration carried out in the major consuming countries themselves, where the potential is much more limited and ultimate costs very much higher than in OPEC countries.

The question which has to be posed now – and urgently – is: Could this situation be expected to continue indefinitely? Could OPEC be reasonably, and under all circumstances, expected to serve as a warehouse, devoid of any will, always being counted upon to fill the role of residual supplier? I, for one, certainly do not think so. If OPEC has not yet reacted in a negative manner to the onesidedness of the consuming countries, this does not mean that a situation like this will persist forever, especially if the circumstances change, as they are very well expected to do. In any game involving two partners, it is natural that the strategies, or actions, of one side will influence the reactions of the other.

First and foremost, one has to consider what general common interests might exist between the consumers and suppliers in the same world in which they live. Obviously, both sides are interested in a smooth and orderly flow of oil supplies, with as few disruptions as possible. At the same time, the oil-producing countries have a deep and direct interest in the growth and stability of the international economy, since any general depression in the world economic activity might affect them, both directly as exporters of oil, and as countries in the process of development. Furthermore, since energy is such an important source for the continued survival of human civilization, both parties should be interested in avoiding sudden shortages, which might lead to conflicts, strife, and possibly chaos.

If we take this argument further, we have to go beyond the area of common interest to both parties and search for the areas of direct interest to each of them.

The OPEC countries in general are developing countries characterized by very heavy dependence on the export of one single raw material. They export oil not for the sake of getting foreign exchange as an end in itself, but rather to use it in the process of enhancing their present welfare and the welfare of their future generations. This means that the best use they can put their resources to is development expenditure in various forms, including the establishment of the necessary infrastructure, the construction of development projects for internal consumption, and, inevitably, the investment in export-oriented projects. Investment in exports is necessary to replace the dependence of their economies on an exhaustible asset with a more viable reliance on more permanent reproduction processes. Therefore, the development efforts of OPEC countries require the active involvement of more advanced consuming countries in the process of technological transfer, development of human resources and know-how, and access for their exports to markets of advanced consuming countries. OPEC countries have been able at extremely exorbitant costs to buy projects and sometimes know-how. However, they are facing severe resistance in their attempt to develop industries which are based on their raw material resource. Refining and petrochemical projects in OPEC countries are facing very serious difficulties and we are hearing threats of protectionist action on the part of potential importers of these export-oriented projects.

As OPEC countries are moving forward towards a more balanced situation of income and expenditure, as their investment absorption capacity improves, it will become increasingly difficult for them to face sudden fluctuations in their incomes. We are gradually moving away from the situation where a number of OPEC countries had current account surpluses which resulted in the ability of some of them to act as buffers for supply-demand fluctuations. This means that in the medium term oil and gas exports of OPEC countries can sustain much less fluctuations than at present, since these countries will need a steadier flow of income. Needless to say, it is only natural to expect this income to be maintained in real terms against inflation and currency fluctuations. Furthermore, this income should be rising in real terms over time to reflect the exhaustible nature of petroleum resources, and to encourage development of alternative resources of energy.

In the short term, the current account surpluses of some member countries, which have arisen as the result of the desire to meet the essential needs of the consuming countries, have to be given sufficient guarantees against various types of risk, including erosion by inflation and currency fluctuations. This is considered by OPEC countries as a fair and legitimate stand, since these assets are mostly recycled back to the major consuming countries in a direct effort to avoid serious disturbances in the financial and capital markets of the world.

In closing, let me say that besides the common interest of all OPEC countries in their relationship with the consumers, there are interests which are common to only some, while, of course, there are areas of interest unique only to a particular country. Without going into any details in analysing these various interests, an obvious step in the right direction on the part of the consuming countries would be to put an end to the present one-sided and selfish attitude towards this relationship. If this could be achieved, then we would look towards a future in which it will be possible for both parties to reach their common objectives. Indeed, a harmonious relationship can only evolve when the interests of both parties are truly taken into account.

CHAPTER 7

# THE US ENERGY SITUATION: AN OPEC
# VIEW

*The following is the substantial text of an address delivered on the
occasion of the Centennial Meeting of American Bar Association,
New York, 8 August 1978.*

The world energy outlook, particularly as regards supply and
demand for oil, is increasingly demonstrating that just as a given
law can be open to interpretation in different ways, so can energy
policy trends. But one should qualify this statement, since while in
law the limits of interpretation are set by the existence of written
legal texts, precedents, conventions and so forth, this is not the
case in the energy field, where the existence of variables creates a
climate of uncertainty in which interpretations of trends thrive
and multiply. Variables such as political will, technology, know-
how, environmental considerations, economic exigency, to name
but a few, make the forecasting of future energy trends a highly
complex task, fraught with pitfalls for the unwary. This has led to
widely differing interpretations of the global energy outlook, and
to disagreement among experts as to the timing and intensity of
what is expected to be the so-called "second energy crisis". As
evidence of this, one has only to note the wealth of literature
published in the last year or so on the subject, and compare the
various conclusions.

The OECD, Exxon, the CIA, the World Energy Conference,
the Workshop on Alternative Energy Strategies, and numerous
independent experts have all applied themselves to the problem
and all, to varying degrees, have forecast a tight oil market
situation developing with disastrous consequences for global oil
supply, which it may, or may not be possible to postpone through
conservation and other measures. The pessimists among them

already predict a doomsday before the end of the 1980s, with a new energy crisis caused by demand outstripping supply, hitting the world. Others, not quite so pessimistic, feel that the crisis cannot be avoided altogether, but that current trends suggest it may be postponed until towards the end of the century. Finally, there are the optimists, who claim that the existing reserves are sufficient to help the world come through the crisis relatively unscathed, or even avert it entirely, especially if national and international policies for the prevention of a tight supply-demand situation are adopted.

Whichever viewpoint one wishes to adopt, one is forced to ponder the question: Why has there been so much interest of late in the future energy outlook? It would be naive for anyone to think that mankind has only suddenly discovered that oil is running short. Perhaps a better explanation can be sought in the decisions and action taken by the Organization of the Petroleum Exporting Countries in the last quarter of 1973. In that year two important events happened on the world scene. Firstly, OPEC decided to fix the prices of its oil unilaterally, resulting in an increase which went some way towards making up for the long-term depreciation of oil prices which it had suffered at the hands of the international oil companies. Secondly, OPEC set in motion a process which has resulted in the ownership of oil reverting to the producing countries. Thus, OPEC contributed to the awareness of the exhaustible nature of oil, together with its adherent and serious implications, thereby bringing sharply into focus the world's energy outlook.

These two events ended many decades of iniquitous treatment and economic exploitation of our countries by the oil companies. Prior to 1973, the production and pricing of oil in OPEC countries was under the total control of the seven major international oil companies. Together, these companies controlled world oil trade, and thus were in a position to keep oil prices artificially low at a time when an industrialized world, recovering from the ravages of war, had based its reconstruction on a cheap energy source – oil. In these circumstances, OPEC countries received a derisory income from sales of their oil of less than one dollar per barrel, while the value of a barrel of petroleum products was worth more than 10 dollars to the final consumer. It was a highly satisfactory state of affairs for the oil-consuming countries, which were able to exert tremendous influence through the oil companies based on their soil. The United States was in a particularly advantageous position, as the home base for five of the top seven companies.

This iniquitous situation had to change, and when the "ancien régime" of the oil world did finally collapse, it was because of the very nature of its shortcomings, and because it was founded on the squandering of cheap oil, regardless of its limited supply and non-renewable nature.

In the United States, the events of 1973 and 1974 assumed a dramatic form, compounded by the psychological impact of the oil embargo imposed by some Arab countries and the increasing reliance of the American economy on imported oil. It came as a great shock to the US citizens to realise that their country, which had historically pursued a policy of self-sufficiency, suddenly had to deal with a group of Third World countries on an equal footing.

Perhaps even more alarming was the fact that in future the United States would have to rely on less developed countries for their energy requirements. In other words, like most of its partners in the West, the United States was unwilling to accept the "fait accompli" that there had been a shift in international economic relations towards the developing countries. It was perhaps too revolutionary for them, and they feared that such action by a group of Third World countries would serve as an example for producers of other vital commodities to follow.

Due to their awareness of the limited base resources of hydrocarbons, and in the face of what they perceived as a threat to their economic well-being, most industrialized countries, including the US, have embarked upon some form of energy policy planning in recent years. Their success has been varied, not least because of the diverse experiences with central planning which different countries had. In the case of the United States, comprehensive and general planning techniques with central guidance from the Federal Government were a concept not generally welcomed, and this sets it apart from most other Western countries. Japan and West Germany, for example, have traditionally promoted centrally-directed cooperation between government and industry, giving rise to guidelines which are generally acceptable to all, and which can be applied smoothly, with a minimum of conflict between the private sector and public authorities. In the United Kingdom, France and Italy, for example, the role of the state is enhanced in the efforts to coordinate energy policies centrally, by means of sophisticated organisations within the government–owned public sector which carry out and coordinate guidelines laid down by the government. In the United States, however, the traditional scepticism about the Federal Government's role in economic policy is compounded by divergencies of interest between Federal and State economic policy. The energy sector is no exception to this rule.

Indeed, it is true to say that teething troubles of this kind seem to be a feature of federal energy programmes in the United States. They began with President Nixon's ill-fated Project Independence which sought energy self-sufficiency for the US by 1980, and were followed up by measures imposed by President Ford in 1975. Neither had much impact on oil imports, as was hoped. Then, in April 1977, President Carter submitted his National Energy Plan to Congress, with its twin objectives of conserving oil and natural gas and developing alternative sources, especially coal. In broad terms, the Plan aims to reduce the annual growth of energy demands to 2 per cent, and reduce oil imports to 6 million barrels per day by 1985 from their 1977 level of around 8.7 million barrels per day and an estimated 12 million barrels per day without conservation by 1985.

OPEC welcomes these objectives, as well as certain other aspects of the Plan, particularly insofar as it favours reliance on a rational pricing policy among the various energy alternatives to achieve its aims. However, this policy seems to have a double standard, which means that rational pricing policy worked only internally, while rational pricing of energy sources including oil in international trade is prevented by various means. On the other hand, serious doubts have been raised as to the prospects of pushing through Congress all aspects of the Plan, and even if it is accepted, there are further doubts about the feasibility of some of its objectives, especially that of reducing oil imports. However, I shall return to this point later.

## US Energy Policy

If we now look at what we really mean by US energy policy, of which the Carter Plan represents the most recent attempt at embodying this into a single legal and administrative instrument, four major areas of emphasis can be discerned. I would now like to examine these in turn and comment on how we, in OPEC, view them.

Firstly, there is energy conservation, which encompasses a wide range of policy options. It has become a watchword of energy policy in all industrialized countries, and yet definitions differ widely as to what we really mean by energy conservation. For our purposes, we will consider conservation in the sense of increased energy efficiency and improved energy transformation – using less energy to produce the same amount or more goods, to put it crudely. It has been estimated that by the year 2020, the global world potential for conservation in energy use would be a saving of about 46 per cent from the level of energy use that might be needed without any conservation measures.

The US has a key role to play here, as the country with the highest per capita energy consumption in the world. A citizen of the US uses up nearly 10 tons of oil equivalent per year, while citizens of France and Switzerland use only up to one-third of this quantity per year. It is also interesting to note that in the US nearly twice as much energy has to be used to produce one unit of GNP (let us say 1,000 dollars worth of products) as is used in France for the same amount of products. These facts throw into sharp relief the capability of the United States' economy if it pursues the right policies, especially as regards the rational pricing of alternative sources of energy and the pricing of energy inputs compared to other factors of production.

As I mentioned earlier, the Carter Plan aims to reduce annual growth in total energy demand to 2 per cent for the period up to 1985. It also aims to reduce gasoline consumption by 10 per cent and to insulate 90 per cent of all American homes and new buildings. It is interesting to note that the targets for annual energy growth rates for the period 1976–85 are much higher for industry than for the residential, commercial or transportation sectors. This probably reflects the heavy reliance of the Plan on conservation policies directed towards persons and households rather than rational pricing policies which have a wider impact on the cost-conscious industrial sector.

It seems that the Plan's targets in the transportation field should not be too difficult to achieve, provided American citizens are prepared to accept certain changes in the lifestyle to which they have long been accustomed. In the building sector, it is felt that the massive investment needed may prove a major stumbling block to the US in achieving its targets, but it is in the industrial sector – where most room for conservation lies – that ironically the US seems most hesitant to act. In 1977, US industry consumed 37 per cent of the nation's energy, and it has been estimated that by 1985 the conservation potential could be as high as 15 per cent. However, it appears that the present Administration is very hesitant to press for measures which might adversely affect economic growth and unemployment.

OPEC welcomes any measures which lead to a more rational use of hydrocarbon resources, provided these are based on the rationale of giving free play to the pricing mechanism. OPEC countries are themselves deeply concerned about conserving their own rapidly depleting oil and gas reserves, lest these should run out before they have laid the foundations of sound industrialized economies.

The second aspect of US energy policy that I would like to consider is its tremendous possibilities for developing indigenous sources. Leaving aside the further development of oil resources, enhanced recovery techniques and the exploitation of shale oil and tar sands, the source singled out for development by the Energy Plan is coal. The Plan calls for a two-thirds increase in coal production to more than one billion tons per year. With nearly 34 per cent of the world's total recoverable coal reserves, the US is obviously in a good position to boost production of an energy source which may become the world's major replacement for oil, as production of the latter declines towards the end of this century, and if the expansion of nuclear power proves unacceptable in the industrialized world. Studies show that due to these vast resources the United States could become a major exporter of coal. If this indeed transpires, then this might go some way towards easing the tight oil market situation which is predicted for the future. However, it remains to be seen if constraints such as legal, environmental and technical problems will hinder optimal development as provided for in the Carter Plan, or indeed if the US will deem it desirable to be a major exporter, given the likely shortage of hydrocarbons.

Nuclear power is the other source which the US may seek to develop further. It is a special case, since although the country has something like 30–40 per cent of the world's uranium resources, depending on its price, we are all aware that the problem is related to questions of safety and the environment, rather than to the technical sphere.

The third aspect of US energy policy, perhaps the one which most affects the member countries of my Organization, is that of lessening its dependence on imported oil, which in practice means OPEC oil, since over two-thirds of oil imported into the US comes from OPEC countries. The United States is not alone in pursuing such a policy, and one can understand the reasoning behind it. However, this desire to lessen the dependence on OPEC oil has been translated into an all-out effort to develop non-OPEC oil resources such as Alaska and the North Sea, often regardless of market conditions and relative investment costs. The impression is that this feverish development is an attempt to offload as much non-OPEC oil onto the market as possible and so weaken OPEC's position. More immediately, such action has led to the current glut on the world oil market, which, although in our view no more than a passing phenomenon, has tended to give a distorted picture of developments in this market.

We, in OPEC, believe that with the continued pursuit of economic growth in the industrialized world leading to increased oil consumption, the demand for oil will rise steadily over the coming years. The world will need all the oil it can lay its hands on, both from OPEC and non-OPEC sources. It is, therefore, with utmost concern that we view this trend towards stepping up exploitation of oil resources outside OPEC, despite generally accepted forecasts by independent experts that more than 40 per cent of ultimately recoverable oil reserves are to be found in our member countries. Moreover, the costs involved in extracting this oil are generally lower than in other areas of the world such as Alaska and the North Sea. However, if global oil and gas shortages are to be averted, the efforts for their exploitation should be more evenly distributed.

The potential supplies from OPEC countries could be enhanced significantly by further exploration and development in addition to the utilization of improved recovery techniques. The estimated potential reserves of crude oil in OPEC member countries of about 550 billion barrels could sustain a production level of 42 million barrels daily between the mid-eighties and the end of the present century. Existing and planned production capacity within OPEC could sustain such a level.

On the other hand, the gas reserves of OPEC countries are in the order of 138 billion barrels of oil equivalent, representing about 40 per cent of total world gas reserves, while the gas production from OPEC member countries is in the order of 4 million bbl/d of oil equivalent, which represents 18 per cent of the total world gas production. It is also important to note that in spite of the low production/reserve ratio for OPEC gas, a considerable amount of its production is flared. Given the right pricing levels and the necessary involvement by the consuming countries, this could easily be put to good use.

However, if we do actually produce such quantities, it will result in our reserves declining dramatically at rates unacceptable to our member countries and it is doubtful whether, by then, such a sharp decline is acceptable to an orderly world economy. Needless to say, some of our member countries will suffer production declines much earlier than the general OPEC level.

Based on the forecasted demand for OPEC oil of about 33.5 and 39.3 MMb/d for 1980 and 1985 respectively (OECD, World Energy Outlook, 1977), it is obvious that if OPEC were to satisfy this projected demand, the time span for exhausting our reserves might not be acceptable in the longer run.

The fourth, and final, aspect of US energy policy which I would like to mention today, is one which encompasses all those measures designed to interfere with the natural forces of supply and demand in the oil market. First and foremost there is the question of stockpiling of oil. All major oil-consuming countries are amassing strategic stockpiles of crude oil and petroleum products over and above the normal level of such stocks held by the oil industry to iron out the seasonal variations in petroleum demand. On the basis of its accumulated experience, the oil industry considers a stock for a period of around 6 weeks reasonable in commercial operations. However, in the last few years, industrialized countries, including the US, have built up strategic stocks (or are in the process of doing so) to levels of 90 days or more. However, the Carter Plan calls for crude oil stocks of 1 billion tons, and this fact, along with, comments expressed in various quarters, seems to suggest that the actual levels of US crude stockpiles are much higher than one is led to suppose. OPEC is concerned that such

stocks held by the US and other member countries of the
International Energy Agency – which together have agreed to
hold 90 days worth of stocks by 1980 – constitute a major threat to
the orderly evolution of the crude oil market and introduce a new,
uncertain variable into the market mechanism despite the cost
involved.

*Interdependence versus isolationism: overcoming the psychological barrier*

In this brief survey of US energy policy I have tried to outline the
major problems as we in OPEC perceive them. However, there is
a purely psychological problem underlying the attitudes of the
United States to energy matters which should not be overlooked,
namely that the US does not yet seem ready, in a world
characterised by interdependence, to accept a degree of reliance
on others, as the rest of the international community has done. For
example, total US foreign trade was less than 10 per cent of its
GNP in 1976 (9.4 per cent imports and 7.9 per cent exports). This
compared with 25 per cent in the case of Germany in the same year
(25.2 per cent imports, 27.5 per cent exports). It would not be the
end of the world for the US to rely to some extent on foreign
supplies, even of a raw material as sensitive as oil. After all, many
countries rely upon the United States for supplies of certain
essential materials. Even a limited export drive on the part of the
United States' economy could easily wipe out the deficit in the
balance of payments which is the cause of most of the anxiety.
Foodstuffs and industrial goods, in particular, could be the key
areas for such an export drive. The main surplus countries of the
OECD, West Germany and Japan, import much higher percentages of their energy requirements than the US, without
crippling their own economies.

What is more, OPEC countries themselves are increasing their
dependence on the industrialized countries, especially the United
States, for imports of goods and technology needed for the
fulfillment of their development plans. OPEC imports from the
OECD area were about three-quarters of the total imports of our
countries. We depend on the United States for about 18 per cent of
our imports, about 12 per cent from West Germany, and over 13
per cent from Japan.

Nor should it be forgotten that our dependence on the US

dollar, as the currency in which oil is traded and in which we hold many of our overseas investments, has led us to expose ourselves to severe financial losses as this currency has experienced fluctuations on exchange markets to our detriment. OPEC will thus be looking to the United States' Government to play a more positive role in stabilising the dollar, without which an orderly evolution of energy supply and demand will be seriously impaired. The United States' efforts in combating inflation at home and abroad will also be carefully watched by our countries, which suffer substantially from imported inflation.

In conclusion, I feel it is my duty to stress my belief that we are not irredeemably locked on a collision course. Nor are we doomed to suffer an energy crisis in the next decade or even this century. On our part, as an organisation and as individual nations, we realise the inevitability of cooperation and interdependence with the industrialized world. We have the resources and the potential to meet all the reasonable requirements for petroleum that help us to make the transition to other alternative technologies. Our countries have shown great consideration for the needs of the industrialized countries for adjustment and recovery. We have done more than any other group to bridge the dangerous gap between the rich North and the poor South.

However, there are several indications that the industrialized countries in general, and the United States in particular, are not really facing up to the challenge of survival. There is an urgent need for a commitment on the part of the United States and the rest of the industrialized oil importing countries to give sufficient consideration to the development needs of OPEC countries. The concern for combating OPEC and complaining about its role will only make the future of both sides gloomier. We should accept the challenge and, in doing so, address ourselves to averting the dangers of the North-South divide, and take the psychological shock of facing up to the realities of interdependence. The United States' economy, more than any other economy in the world, has the dynamism to steer clear of the crisis if action is taken now.

*Postscript*

This paper was written in 1978 and it commented on US energy policy as formulated by President Carter in his 1977 Energy Plan.

There were, however, important developments in US energy policy after 1978. President Carter, at the time of the Iranian revolution, introduced a programme of gradual de-regulation of domestic oil prices in the US. Prices were allowed to rise at different rates depending on whether oil was classified as "old" or "new", and on whether oil was obtained from stripper wells or through enhanced recovery technique. The transition period leading to full de-regulation involved price distortion, new taxes intended to cream-off windfall profit and new control and allocation schemes.

President Reagan brought forward the end of price regulation on domestic oil by several months. He dismantled the burdensome system of control, regional allocation and import entitlement which accompanied price regulation. President Reagan also changed the direction of US policy in the field of energy substitution, largely for ideological reasons. The Republican administration is limiting the contribution of the Federal Government towards the development of synfuel and other energy sources. The new philosophy is that such investment should be left for the private sector – not undertaken by government nor encouraged with larger Federal subsidies. President Reagan is also determined to relax the environmental and other regulations which have so far hindered investment in coal and nuclear in the US. In short, the new policy emphasis the role of market forces, and attempts to remove as many regulations as possible. The US Government, however, will continue to play a role in matters of national security. It intends to build up the strategic petroleum reserves, the 750 million barrels target is reached probably by the end of this decade. The US dependence on imported oil has substantially reduced in the past two years. But the apparent success should not entirely be attributed to US energy policy, as it is largely due to the economic crisis which is affecting all developed consuming countries.

# CHAPTER 8

# THE PRICING OF PETROLEUM

*The following is the substantial text of an address delivered on the occasion of a joint Seminar sponsored by OAPEC and the Norwegian Petroleum Society, Oslo, 29 September, 1978.*

## Introduction

It is indeed a pleasure for me to address this Seminar on the vital issue of petroleum pricing, particularly in view of its co-sponsorship by OAPEC, our sister organization with whom OPEC shares seven member countries. I am also pleased to note that this Seminar is being held in Oslo, since as an oil-producing country, Norway has many things in common with OPEC members. Although at the same time, it has to be acknowledged that, being a developed country, Norway's position in the world economy also differs in many ways from that of OPEC countries. However, I believe that cooperation between Scandinavian and OAPEC countries, which is the theme of this Seminar, can, if pursued, lay the foundations for much needed cooperation between the industrialized countries of the North and the developing countries of the South. The Scandinavian countries are well-fitted for this role, since their thinking on development matters is not clouded or inhibited by the outdated influences of the colonial past, as is the case with many industrialized countries. Furthermore, their comprehension of the current problems faced by developing countries enables them to play an important role in reconciling North/South differences.

## The nature of oil prices

In my presentation here today, I would like to try to unlock some of the mysteries surrounding price formation in the oil industry, and to give some indication of how we in OPEC view future trends in this area. The pricing of petroleum is indeed a subject about which much has been written and even more discussion held. However, the tools which economists normally use to provide insight into the pricing of goods have proved insufficient in the case of oil.

We will never find the answer to the evolution of oil prices over the last 20 years by looking at marginal costs, for example. We can only find it in an objective analysis of the forces which have a vested interest in oil prices and which are able to exert influence over them. These forces have been able to bring pressures to bear on prices, thereby manipulating them in their own interest.[1]

Prior to 1973, the concept of "price" in the context of international oil trade had become notorious for its misuse rather than for its usefulness. Right up to the early 1970s, this trade was conducted through the integrated channels of the Majors, so that oil was produced, transported and marketed by the same company through these channels. Even when the oil multinationals traded oil with each other, little information was published on the deals. In short, crude oil prices at that time were no more than entries in the books of the oil companies. What were generally known as "posted prices" were no more than useful reference numbers for calculating how much tax the host country would be allowed to levy.

Changes in the price of oil in the pre-OPEC days were never a reflection of the evolution of production costs or that of alternative energy sources, but rather a reflection of a shift in the balance of forces making up the "controlling power". Even after the creation of OPEC, which was able gradually to chisel away at the power base of the "controlling power" and half the decline of prices, the oil companies fought tooth and nail to preserve their sole right to decide oil prices.

Nor should we forget that while OPEC was fighting to raise the revenue earned by one barrel of crude in the 1960s from the figure of less than 1 dollar, the oil companies were selling the oil in the form of petroleum products for about 12-14 dollars per barrel in Europe. Indeed, when we talk of oil prices, it is more meaningful to consider such product prices, since they have direct utility for

[1]For a detailed analysis of this issue, see "The Pricing of Oil: Role of the Controlling Power", a lecture delivered by Ali M. Jaidah at an Energy Seminar sponsored by Harvard University, Cambridge, Massachusetts, May 9th, 1977.

the consumer, while crude does not. It is, of course, still a fact today, that petroleum product prices reflect a greater share of revenues for consuming governments (in the shape of taxes) than they do for the producing governments.

## Factors influencing price making

The significant changes in the structure of the petroleum industry in late 1973 and early 1974, the situation which I have just outlined to you, has undergone considerable transformation in two important areas. Firstly, as from October 1973, OPEC took over the price-fixing function for crude oil, and this was never again to be the subject of negotiation with the oil companies. Secondly, the OPEC countries set in motion a chain of events which has led to their assuming full control over upstream activities of the oil industry through nationalisation and participation, and thus the producing countries replaced the oil companies in determining the production volumes for crude oil. The combined effect of these two factors has been to add a new dimension to the current oil scene which has radically altered the face of price formation in the industry. I would now like to say a few words on what we understand by this new dimension.

In the first instance, the price paid for oil is in reality the revenue of the producing countries' governments. For business, prices are set within the normal economic criteria of the structure of the market, the marginal revenue equated with marginal cost, and in the end the rate of return. While for developing, oil-exporting countries, the price received for their oil is the predominant source of their revenue, if not the only one, and certainly the only source of foreign exchange earnings. The price received, therefore, is not dividends to be distributed to a few shareholders, but rather directly affects the lives of all inhabitants of those countries which are dependent upon it. The whole national prosperity of OPEC countries is dependent on revenues received from oil sales. From this source, schools and hospitals are built, roads are constructed, electricity is generated, development projects financed and for many, sea water desalinated. With oil being by nature non-renewable, and with the horizon of its depletion approaching, our member countries are in a race against time to bring their economies up to the stage of sustainable growth and maturity. We are repeatedly faced with the question as to what will happen to us after the oil era. Having adapted ourselves to some degree of sophistication of modern life, we are not prepared, nor eager, to go back to those "good oil days" with all the niceties of primitive life.

Having said that the price of oil is in essence the revenue of our member countries, OPEC Ministers, when deciding OPEC prices, would ultimately be concerned about maintaining the real purchasing power of their income, if not actually increasing it. There are two interrelated factors affecting our purchasing power, namely that of inflation and exchange rate fluctuations.

The rising prices of the goods which OPEC countries import from the industrialized countries, coupled with the effects of the sinking dollar on their oil incomes, have interacted to cut the real value of the revenues which our countries derive from their oil exports. Indeed, it has been estimated that the real value of a barrel of our crude has been more than halved in terms of 1973 dollars. All OPEC countries have suffered losses of revenue, therefore, particularly as the price of oil has been frozen since December 1976. The current account surpluses of OPEC countries, which experts forecast in 1973 would have now have gone through the roof, have dwindled, and a number of countries are now net borrowers on the international market. The IMF estimates that the OPEC current account surplus this year will be just 18 billion dollars (67.8 billion in 1974), and if this trend continues, it will shortly disappear altogether. Meanwhile, total OPEC borrowing in the form of Eurocurrency bank credits and international bond issues has already risen to 5.6 billion dollars in the first half of this year. OPEC countries are all in the process of implementing economic development plans designed to transform their economies into industrial entities able to stand on their own feet when the oil runs out, as it must inevitably do one day. They are fuelling their development with the one resource they have: petroleum. Falling real revenues can, therefore, only lead to a call among OPEC countries for increasing crude prices.

In the second instance, unlike other commodities, oil is and always was of strategic importance. As early as World War I, Georges Clemenceau said that "one drop of oil is worth one drop of blood" and Lord Curzon confirmed this. "Truly", he said, "posterity will say that the Allies floated to victory on a wave of oil". The Arab oil embargo of late 1973 illustrated this reality and its impact was felt in every household in the industrialized countries. Moreover, the increase in the price of crude oil in late 1973 drastically changed its impact on international trade and thus on the balance of payments of the importing countries. Therefore, the process of price-fixing has to be considered amongst other things in the context of political and strategic considerations. The consuming countries' governments, like any other consumer, have always wished it to be carried out in such a way that they can buy as cheaply as possible.

The third aspect of recent changes in the oil industry affecting oil price formation is that related to supply. Whereas the oil Majors used to coordinate their liftings of crudes from many sources in accordance with the logistics of downstream demand requirements, oil production is now determined by each OPEC country in accordance with its national economic imperatives and political strategies, many of these countries having a limited type of crude.

Furthermore, an ever-growing volume of oil is being sold on the market by the National Oil Companies of the OPEC countries. Indeed it is estimated that about 25 per cent of total OPEC production entering trade, does so through this channel. In the absence of central coordination and programming of production levels, and in the presence of an assumed over-supply situation, the pricing differentials of various crudes are becoming of great concern. Any price disadvantage for one crude may affect substantially its level of production and in the case of a country with one type of crude, such as Kuwait, this is detrimental to its interests. We refer to this in OPEC as the relative value problem, which I shall return to later.

Fourthly, the increasing role of OPEC and its member countries in the oil industry has brought about a reaction from the consuming countries. The industrialized countries are becoming more involved in energy matters. Policies have been formulated to diversify and promote other sources of energy including non-OPEC oil. Under these circumstances, OPEC is increasingly being made to bear the brunt of supply fluctuations in its role as "residual supplier". In plain terms, oil consumers are seeking to maximize their independence from OPEC oil by developing as many indigenous and non-OPEC oil and other energy sources as possible. This policy has the effect of compounding all the small changes in total world energy supply and demand, accumulating them, and finally reflecting them as considerable fluctuations in OPEC exports. This resulted in what we have experienced over the last year, a glutted oil market which has exerted a downward pressure on crude prices. This role of residual supplier, which the consuming countries would have us play, is definitely an unwelcome one for OPEC countries, since it inevitably leads to fluctuations in revenues.

In the past this has worked because certain OPEC countries, notably Saudi Arabia, Kuwait, Venezuela and the Libyan Arab Jamahiriya were happy to absorb the bulk of production cutbacks. However, this situation is changing as government expenditure commitments are increasing and production levels are approaching the desired minimum level for those countries.

### The OPEC approach to oil pricing

What I have outlined above are the key issues determining the price of crude oil exports. I would now like to address myself to the ways and means by which OPEC, and in particular its Secretariat, analyse those issues with a view to improving comprehension.

Firstly, OPEC is closely following the impact of inflation imported into our member countries, based on the pattern of our imports. We compare this with the rate of domestic inflation of export prices reported by major industrial countries. Additionally, we monitor the developments in the field of exchange rates. The effects of these parameters on the purchasing power of our revenues is then evaluated.

In conjunction with another factor influencing OPEC revenues, namely the demand for our oil, we are closely following the level of economic activity and the various energy strategies (conservation measures, stockpiling, etc.) that might affect this demand in the major industrial oil-importing countries. Coupled with this, we also pay close attention to the flow of non-OPEC oil supplies. On the basis of these and other factors, the supply-demand outlook for the short term is finally projected and evaluated.

Secondly, OPEC is studying the values of OPEC crudes relative to the price of the Marker Crude (Arabian Light). The idea is to produce a coherent price structure for OPEC crudes whereby transport and quality differences are allowed for in such a way that the crudes have a price equalling their relative value to the refiner at the point of use. With such a price structure, the incentive for consumers to substitute between the individual crudes is reduced and no one crude will be adversely affected when a short-term shift in demand for OPEC oil occurs.

Various approaches for determining relative values are used by the Organization, including net-back costing, the replacement value method and simulation of a world-wide competitive market for all crudes.

The f.o.b. value for each crude in the net-back approach is obtained by subtracting from the composite value of the barrel derived from retail prices, the costs of distribution, refining, transportation, duties, and respective profits. The export pattern can then be used to provide a weighting factor so as to arrive at a single value. What hampers the use of this method is the difficulty in obtaining the enormous amount of reliable and up-to-the-minute data required.

In the replacement value method, the gross product worth (g.p.w.) of each crude is obtained by adding up the values of the refined products using spot market prices and assuming a representative refinery configuration. Some quality adjustment is then applied to cater for those products with characteristics outside the market specification. The cost of transportation is then subtracted from the adjusted g.p.w. to arrive at the f.o.b. value, which is then compared with that of the Marker Crude to obtain the differential. Finally, the weighted average differential can be generated using the export pattern. One difficulty in applying this method lies in estimating the quality adjustment to meet market specifications.

The concept of the third approach is to satisfy a given crude demand pattern by a given crude slate at a minimum cost. This is done by optimizing world-wide crude flows during a given period, assuming a state of perfect competition. This involves using a large scale cost-based linear programming model. The dual value of the crude availability constraints from this model represents values attributable to an extra unit of that crude on the market under optimal allocation. Since these values are measured with respect to the Marker Crude price, a set of ideal equilibrium values is produced.

One disadvantage of the linear programming approach is that an accurate technical representation of transportation, refining and product distribution, adequately disaggregated over crudes and regions, implies a very large mathematical programming problem, and considerable human and computing resources are needed to run a project on this scale. Even so, there are various political and other relevant real life considerations that can never be properly represented in the model.

Useful as these scientific methods may be in facilitating comprehension of the issues involved, it is very difficult to conceive an approach that would simulate the actual realities or which would be fair and acceptable to all member countries. Therefore, an element of compromise is always left to the wisdom of the price-setters.

Finally, OPEC is studying what has come to be known as the absolute value of crude oil. This is important for the long-term pricing strategy of OPEC, a problem which comes within the mandate of a special OPEC Committee of six Ministers (from Algeria, Iran, Iraq, Kuwait, Saudi Arabia and Venezuela), set up after the Informal Meeting of OPEC Oil Ministers in Taif in May 1978. Of course, the brief of the Ministerial Committee on Strategies is to examine overall policies and strategies which OPEC might follow in the medium and longer term. More "technical" research into this highly complex problem is being undertaken by the OPEC Secretariat, including the development of two major long-range modelling projects, both having basically the same objectives but differing in detailed implementation. Both tackle the issue from the global context in which OPEC functions to evaluate how various economic, political and social factors influence oil prices. The principal objective behind our modelling activities is to evolve the optimum strategy for the determination of the price of crude oil and evaluate the impact of pricing decisions on both energy and non-energy sectors of the world economy, including the demand for OPEC crude oil, through consideration of the different feedback mechanisms involved. By working through various scenarios on price or production, it is possible to find a strategy which, at least, approximates the optimal policy for OPEC in terms of an objective, such as maximizing the discounted oil revenue to OPEC, or which satisfies some other suitably defined objective. An alternative would be to market equilibrium pricing for all types of energy, and yet another possibility would be some other imposed pricing pattern.

In each case we need to know the effect on OPEC hydrocarbon exports, on revenue generated and the overall impact on producing and consuming countries' economies. A further set of cases would relate to alternative production strategies, either alone or linked to various pricing strategies.

This is briefly how I see the factors influencing the pricing of petroleum and how OPEC is seeking a better understanding of these factors with a view to rationalising pricing decisions. I hope that what I have said, has at least thrown some light on an issue that is still heavily prone to misunderstanding.

## CHAPTER 9

# PROSPECTS FOR GAS PRICES AND THE DEVELOPMENT OF THE NATURAL GAS INDUSTRY IN QATAR

*The following is the text of a paper published in "Issues in Development: The Arab Gulf States", edited by May Ziwar-Daftari and published in 1980 by MD Research and Services Ltd.*

The potential for growth in the world's natural gas industry and market is immense, for as an industry it is still virtually in its infancy. Its major future significance in the world energy picture will be as a bridge in the transition from oil to other alternative forms of energy. Natural gas has only existed as an important component in the world energy balance over the past three decades, for until the 1950s its share was a negligible 10 per cent (except in the US, where it accounted for about 25 per cent of total energy consumption). Today gas accounts for 18 per cent of the world energy balance compared to 46 per cent for oil, 25 per cent for coal and 8 per cent for other energy forms.

Natural gas's potential share of the world energy balance will continue to grow in the future if only because of its premium quality as a clean fuel. Huge proven and potential reserves to meet expected future demand already exist and are much more important than many people think. In terms of proven reserves, natural gas with 2,250 trillion cu. ft (t.c.f.) (or 450 billion barrels oil equivalent (b.o.e.)) accounts for at least 8 per cent of total world proven reserves of fossil fuels. In terms of its potential or ultimate reserves, natural gas has even better prospects which are estimated as being equal to ultimate world oil reserves put at 10,000 t.c.f. or 2,000 billion b.o.e. Deeper drilling is expected to result eventually in more gas discoveries. As it is, currently the ratio of proven to ultimate reserves is about 30 per cent for gas as against 60 per cent for oil. Carrying the comparison with oil further, proven remaining gas reserves have doubled over the last 10 years, while remaining oil reserves have only marginally increased. In addition, world oil reserves at present consumption rates are expected to last 25 years, while known gas reserves will last 45 years. In fact the natural gas life index will rise substantially because gas discoveries have so far been largely accidental.

Interestingly, the geographical distribution of world gas reserves and consumption shows the same imbalance between supply and demand as with crude oil. For instance the USA, the world's major oil consumer, also consumes about 40 per cent of world marketed gas, yet it only accounts for 8 per cent of reserves. Japan imports all the gas it consumes, and Western Europe buys 13 per cent of marketed gas with reserves put at 6 per cent. In contrast OPEC's share of world reserves is estimated at 40 per cent while its share of consumption is only 6 per cent. In the USSR and other COMECON countries gas reserves seem capable of meeting growing demand.

When compared to crude oil, internationally traded gas is much smaller, accounting for only about 12 per cent of total consumption about 6 t.c.f. (about 3 MMb/d of oil equivalent). While about 50 per cent of world crude oil output is traded internationally, about 80 per cent of the gas traded internationally is transported via pipeline, mostly in interregional transfers. For the rest, about 1.2 t.c.f. (about 600,000 b/d oil equivalent) is moved in LNG form, exclusively for intercontinental transfers.

The world's top liquefied natural gas exporter is Algeria with 1.3 billion cu. ft/day (0.5 t.c.f./year) followed by Indonesia and Brunei, each with 0.75 billion cu. ft/day (0.25 t.c.f./year). The rest of the world gas trade is accounted for by exports from the Libyan Jamahiriya, Abu Dhabi and Alaska. At the receiving end, Japan holds top position, consuming about 1.8 billion cu. ft/day (0.6 t.c.f./year) from Brunei, Indonesia, Alaska and Abu Dhabi. Europe imports 1.1 billion cu. ft/day (0.4 t.c.f./year) from Algeria and the Libyan Jamahiriya, while US imports from Algeria amount to 0.6 billion cu. ft/day (0.2 t.c.f./year).

As already mentioned, there is obviously a tremendous growth potential in the international gas trade. Projections of the natural gas supply-demand outlook for the major consuming importing countries indicate a large gas supply deficit developing towards the end of this decade which will have to be met from massive intercontinental transfers. In Western Europe, added production from the North Sea is unlikely to be sufficient to counter a regional decline in production. The region is expected to need an extra 5 billion cu. ft/day of gas imports above the volume already contracted for, on the basis of a modest growth in consumption of 3 per cent a year (as against an historic growth rate of 20 per cent over the past 10 years). In the US output is also expected to decline, and even taking into account increased imports from Canada and Mexico, and large movements from Alaska as well as some contribution from synthetic gas, there is no margin for growth in demand at current levels beyond 1985. So, unless significant imports are committed, supply limitations on gas consumption will be felt in the US by the second half of this decade. In Japan MITI is already planning for massive LNG imports, increasing from the present 2 billion cu. ft/day to about 6 billion cu. ft/day by 1990.

Even if the gas share in the world energy balance remains at its present level of 18 per cent, world gas consumption, running now at 50 t.c.f. is likely to reach 80 t.c.f. by 1990. Most of the increase in the US, Europe and Japan, will have to be met from imports. A large number of gas projects has been planned to meet this scenario, and if they materialize world trade could increase two-fold by 1985 and three-fold by 1990. The most important markets for new LNG supplies are likely to be Japan and developing countries. Pipeline exports are expected to continue to account for the highest share, moving from the present 5 t.c.f./year to about 11 t.c.f./year. But LNG imports are expected to increase even more sharply from the present level of 1.2 t.c.f./year to about 6 t.c.f./year by 1990.

Qatar now finds itself in a particularly favourable position to take advantage of the expected huge growth of the world natural gas trade and gain a significant share in the LNG area. Qatar's natural gas potential that has so far been discovered, is impressive by any standards. Gas in significant amounts was only discovered in the 1970s and so far has been put, at a conservative estimate, at 150 t.c.f. The full extent of its gas reserves has yet to be determined, as at this stage only six gas exploratory wells have been sunk. But the extent of the gas reserves already determined, in terms of BTUs, is almost six times the size of Qatar's oil reserves. Thus the prospects of LNG exports for Qatar present great attractions as a significant contribution towards its economic and general social development, as well as providing an important alternative source of revenue to its oil earnings.

The immense size of Qatar's gas potential will place the country well among the world's leaders in terms of gas reserves. Discovery of this huge gas resource comes at a most opportune moment for its development will complement the limited nature of Qatar's oil resources. The present and future availability of this natural gas, along with the industrialization processes involved in its development and utilization will open up new horizons for Qatar, particularly in the field of LNG exports. As already mentioned it will enable Qatar to widen and diversify its sources of revenues by making it far less dependent on its oil resources – an important consideration for a country with few other natural assets. A further attractive feature is that this gas will sustain the available energy resources for Qatar into the "after-the oil era". The implication in the medium term is that more of Qatari crude oil that would otherwise be locally consumed shall be available for export, and in the long run it will not depend on imported energy.

More indirect benefits will involve its use in expanding Qatar's water resources and purification, particularly in the use of gas to power desalination projects. A spin-off from such water development may be to improve agricultural potentials in the peninsula. Natural gas, by its nature, also opens up opportunities to develop a wide range of industrial projects, which are being currently carefully evaluated, whether as a valuable source of energy in energy intensive processes such as aluminium and steel smelting, or as a feedstock for the highly complex production processes of the petrochemical industry. In general, therefore, the existence and careful exploitation of these gas resources in the future can only enhance the potential economic development of Qatar.

A basic parameter for gas utilization in Qatar or elsewhere will of course depend in the first place on the cost of production because of the highly capital intensive nature of the industry needed to exploit the resource. Where Qatar and other Gulf countries are concerned there is also the lack of a nearby consuming market. Thus the transportation of gas as LNG will form a high proportion of the CIF price. Much, therefore, will depend on the price consumers will have to pay to make exploitation of this valuable form of energy feasible from the point of view of gas producers. Consumers of gas will have to come to accept a price mechanism that takes account of the particular attractive properties of natural gas.

So far the slow growth of the international LNG trade is partly explained by the hesitation in the past of some producing country governments to press ahead with export projects. One hindrance was that until very recently imported natural gas prices were set by consuming countries in reference to the price of residue or No. 6 fuel imported into these countries. Domestic gas prices in these consuming countries were also held down to levels frequently below alternative energy prices, as is the case still for some US gas. In addition, as already indicated, the high cost of LNG transportation (on average about five times that of oil on a BTU basis) and the high costs of LNG liquefaction plant and manufacture, along with low netted back value from exported gas represent only a small fraction of the value derived from crude oil exports, and provide little incentive to producing governments in the Gulf area.

But recently prospects for the gas trade have undergone some improvement, through increased oil prices registered since last year and the growing awareness of an approaching energy crisis. This awareness was brought home to consuming countries through the temporary energy shortages some experienced last year in the wake of the Iranian Revolution. As a result some consuming countries now seem to be coming round to accepting the concept of direct parity between CIF gas prices and oil prices on a BTU basis.

But even this, though a significant advance for gas producers, still means a netted back value for gas at the wellhead far below that derived from oil. For example, in the case of exports to Japan from a typical Gulf LNG project, based on a parity with an average landed price of $30 per barrel of crude, it would result in a CIF value of LNG of about $5.25 MMBTU. This would translate into a f.o.b. value of about $3.75 MMBTU and a netted back value at the wellhead anywhere between $1.50 and $2.50 MMBTU depending on the costs involved in the producing countries, for production, gathering, transportation and liquefaction of the gas. This range translates into a wellhead value of $8 to $14 per barrel of crude oil on a BTU basis.

If parity with crude oil is sought on a f.o.b. rather than a CIF basis, the netted back value at the wellhead will be improved by the difference between transporting LNG and crude oil from the Gulf to Japan based on the above calculation. The difference would be about $1.25 MMBTU, raising the wellhead value of gas to between $2.75 and $3.75 MMBTU range, equivalent to $15 to $21 per barrel of crude oil on a BTU basis.

While this approach might give some producers sufficient incentives for investing in gas production, there are further considerations to bear in mind. Natural gas should not be viewed as in competition with crude oil, but rather with oil products such as Fuel No. 2, No. 6 and possibly with naphtha as a feedstock for petrochemical production. The pricing mechanism, therefore, should also take account of gas being a clean flexible premium fuel on parity with the "noble" uses such as those for which gas oil and naphtha are presently reserved.

In an energy market environment where supply, particularly of crude oil, is unable to meet demand, the pricing of gas, as a supplementary source of energy, should possibly not be calculated on its replacement costs with oil. Such replacement value concepts, it can be argued, only work in a market when supply of gas is intended to compete with crude oil or its products in order to replace it. But in the reality demand for natural gas is to supplement the physical limitation of crude availability. From this perspective some gas-producing countries may wish to achieve full parity at the wellhead between natural gas and crude oil on a direct BTU replacement basis.

Perhaps the only long-term reference that in the final analysis will satisfy everyone will be linking the price of natural gas with its only clear alternative, coal gasification or synthetic gas. This is the only alternative capable of providing abundant fuel of the same quality as natural gas and it cannot be supplied at a cost below $8 in the major consuming countries. On this comparison, the f.o.b. yield for LNG would be about $6.50 MMBTU with a wellhead value of $5, which would bring it to the equivalent of revenues derived from oil at the wellhead.

On this basis, imported natural gas in the consuming countries would be priced well above the value of oil or any of its products. But there are good reasons why gas should be given such a preferential treatment. As already mentioned, it is a clean burning fuel with considerable environmental benefits and it can be used as a valuable feedstock in the petrochemical industry. For practical reasons it also has to be sold on a long-term contractual basis, which for consumers means a guaranteed availability for 20 years. In addition importing countries can expect considerable savings in foreign exchange, as a substantial portion of the CIF price could be returned in payments for plants, equipment, shipping and other transportation charges. Such a pricing approach would provide a real incentive for consuming countries to conserve gas for its "noble" uses and provide a further incentive for the development of more expensive alternative energy sources, particularly, synthetic gas production. In the long-term the adoption of such principles in the pricing mechanism is the only solution to providing the world community with enough energy to meet its growing requirements.

# CHAPTER 10

# EVALUATION OF OPEC POLICY
# RESPONSES TO THE ENERGY PROBLEM
# SINCE 1973

*The following is the substantial text of a lecture delivered at the Second Oxford Energy Seminar at St. Catherine's College, Oxford University, England, 8 September, 1980.*

The Organization of Petroleum Exporting Countries is celebrating its twentieth anniversary this year. This anniversary will be marked by a number of events: a special Conference of Ministers of Oil, Finance and Foreign Affairs, an important seminar in Vienna on OPEC Long Term Strategies and above all a Summit meeting of the Heads of State of OPEC member countries early November in Baghdad. Anniversaries are privileged occasions for an examination of the achievements made and the difficulties encountered in the years that have passed; also for a fresh look at the challenges, and the obstacles that may lie ahead in the years to come.

The invitation to speak to this distinguished audience provides me with a fitting opportunity to engage in such an exercise. I would like to consider major issues and significant developments which have illustrated the history of OPEC in recent years. I do not, however, intend to indulge in romantic reminiscences of important and sometime dramatic events. My purpose, rather, is to analyse as objectively as possible relevant aspects of the past in order to draw hard and useful lessons for the future.

By necessity, the approach followed in this paper will be highly selective. It is impossible to undertake in a short speech an exhaustive study of the issues raised by the eventful history of OPEC and to relate these issues to the tasks, opportunities and problems likely to face the Organization and its member countries in the next ten or fifteen years. I propose therefore to concentrate on three major themes. The first theme is petroleum prices, or, more precisely, the function exercised by OPEC in administering the price of oil. The second theme relates to the significant structural changes which have taken place in the marketing of crude oil by OPEC countries. Finally, the third theme is that of OPEC's international image and role, a theme which involves a consideration of new relationships between OPEC and the industrialized world and OPEC and other developing countries.

*Prices*

That the oil price revolution of late 1973 was the major event of OPEC's recent history is a fact that nobody disputes. The significance of this event was twofold. It involved, first, a sudden and dramatic adjustment of the price of oil; and, secondly, a transformation of the nature and role of OPEC.

An oil price adjustment in 1973 was in a sense unavoidable. Though it shocked the world, the adjustment was both necessary and long overdue. It reflected a fundamental change in the perceptions of the industry about the scarcity of oil, a depletable commodity whose consumption was growing at a faster rate than that of new discoveries. The adjustments would have been less disruptive for the world economy had it been possible to introduce them earlier in a gradual manner. But OPEC, in the years preceding 1973, was a relatively weak organization and its ability to determine prices was extremely limited. It was not in a position to make the gradual price adjustments that were necessary since the early 1960s. The companies which were vested with the "controlling power" over oil prices were not concerned with long-term scarcity. They were more involved with the short-term conditions of the oil market and were engaged actively in expanding their marketing outlets. Long term considerations called for price increases but short-term interest and consideration resulted instead in a lowering of the real price of oil.

In 1973 OPEC succeeded in asserting the full sovereign rights of its members over the disposal and pricing of oil. In doing so it fulfilled to a large extent one of the main objectives of its foundation: to create the favourable conditions which will enable member states to derive substantial economic advantages from oil production and exports. The exercise of full sovereignty in the determination of oil prices is one of these essential conditions. It superseded much of what OPEC was engaged in before bargaining with companies over tax rates and tax formulae in order to maintain the revenues of members. The need to protect interests through protracted negotiations and defensive means had receded, and the opportunity for an active promotion of economic objectives had now emerged.

Pricing related issues continued to be the main preoccupation for OPEC after 1973. One difference with the past, of course, is that the post-1973 OPEC is in a much better position to do so. But there is a second major difference which initially passed unnoticed. OPEC reclaimed its sovereign rights over oil at the time when perceptions about the long-term scarcity of this commodity had suddenly and dramatically changed. OPEC became the administrator of the price of oil when the world began to perceive the threat of an energy supply crisis. This situation has vested OPEC with an important but potentially troublesome international responsibility. The administration of the oil price cannot be exclusively determined by the narrow economic interests of a parochial producers' association. Price administration will necessarily involve considerations pertaining to the management of the energy crisis especially if the administrator believes, as OPEC undoubtedly does, in the seriousness of the problem.

Two main aspects of the energy crisis bear on the pricing issue. In the short-run, the energy crisis coincided with and seems to be associated with an economic crisis, a situation which calls for restraint in pricing behaviour. But the long-run solution of the energy crisis involves gradual, nevertheless substantial rises in the price of oil. OPEC, therefore, has a complex task to perform as price administrator. Prices must be determined in ways which, *first*, satisfy the legitimate revenue objectives of members; *secondly*, signal to myopic consumers, usually over-impressed by the short-term conditions of the market, a correct measure of the relative scarcity of oil in the long term; *thirdly*, do not inflict damaging and long lasting strains on the world economy.

In short, OPEC has become after 1973 the price administrator of a vital commodity affected by a long-term supply problem. To administer prices in this context is not a simple exercise in revenue maximization for a small group of producers. Price administration involves a role in the management of the energy crisis. OPEC has an international responsibility which it cannot evade; to help through appropriate pricing policies the necessary transition from the oil era to a new technological age in which greater reliance is placed on non-oil energy sources. This transition as we all know is likely to be long and difficult and the costs of failure are high.

OPEC's role in pricing is thus complex. It can involve conflict of objectives. To achieve maximum advantages for Members to avoid disruptive shocks and to help contribute to a smooth transition from oil to alternative energy may lead to inconsistencies. The oil companies did not face the same problem when they were vested with the controlling power before 1973 because they did not entertain these serious worries about the long term scarcity of oil. They enjoyed administering the price of oil without having to manage the energy crisis.

Let us now briefly review oil-pricing policies and price behaviour since 1973 in order to measure the degree of OPEC success in fulfilling its objectives and to assess the difficulties and obstacles encountered. It is important for this analysis to distinguish sharply two sets of oil prices. The first set is comprised of a single price, that of Arabian Light 34°API, the Saudi marker crude which serves as a reference to all internationally traded oil. The second set is comprised of the prices of all other crudes which differ in terms of product yield, sulphur content and location. The first price, that of the marker crude, in a fundamental sense, is the OPEC administered price. It can be fully administered irrespective of market conditions which means that this price can be raised even if the supply-demand balance is unfavourable to producers. A common and widespread fallacy is that the marker price cannot (or should not) be increased in time of glut. History disproves this fallacy. Government revenue per barrel of Arabian Light increased to $10.14 in November 1974 (from a previous level of $9.41 in January 1974) despite the marked weakening of the oil market in the latter part of 1974. Further the marker price was raised by 10 per cent in October 1975 (Vienna meeting) at a time of relative glut. A further 10 per cent increase was decreed in December 1977 (Doha) despite adverse demand conditions. And it is fairly safe to predict that the marker price will be adjusted soon above its current level although the market is characterized by excess supplies.

Prices in the second set are administered by the producing country of each relevant crude but administration is strongly influenced by two factors: first, the reference price of the marker crude and second the state of the supply-demand balance. If there is excess supply of oil, prices of the different varieties of crude oil tend to cluster around the marker. Price differentials between good quality crudes and the marker tend to narrow and the negative differentials between heavy crude and the marker tend to widen. This phenomenon was clearly observed in 1974. In the second half of 1974 demand for oil became slack and price differentials began to narrow significantly. Producer countries other than Saudi Arabia had to lower the asking price for their own varieties of crude. Failure to do so with sufficient speed and down to the appropriate level led to considerable reduction in offtake. While the aggregate reduction in demand for OPEC oil in December 1974 compared with September 1973 was of the order of 13.5 per cent, the reduction in Libyan exports was as large as 56 per cent. In January 1975 Abu Dhabi suffered a reduction of exports of 32 per cent in a single month. Differentials were reduced in order to re-establish a proportionate distribution of the overall decline in demand between producers. In one case the price of a given variety of crude was reduced from $16.00 (in January 1974) to $11.86 (January 1975).

Market forces have a powerful influence on the behaviour of price differentials (but not, let me stress again on the price of the marker crude) both in periods of glut and in periods of shortage. When there is excess demand for oil as in late 1973/early 1974 and throughout 1979 producers are able to raise their asking prices virtually as they please. The reason is simple. Because oil, at present, has no immediate substitutes in certain uses there is no economic lid on its short term price in times of short supply. In such a situation individual producers acting autonomously could set widely different prices for very similar varieties of crude and find that they can sell all their available supplies. In a tight market, disorderly pricing does not involve the penalty of a loss of output for the producer whose price is out of line. There is a penalty in the form of foregone revenues for those who fail to raise their prices fast and far. Hence, the chaotic pattern which puzzled observers so much in 1979; the price structure was in chaos and every attempt to bring order to the price structure inevitably led to leapfrogging.

It may appear, at first sight, that the problem of price differentials is more serious for OPEC when the market is slack (i.e. in times of so-called glut) than when the market is tight. The narrowing of price differentials during the glut has a smack of price competition between members of a producers' association which pledged solidarity. When the necessary adjustments of price differentials are substantial strains may arise. I would like to stress, however, that the problems of price differentials during the glut have their origin in the preceding phase of the cycle, in the shortages which encourage a free-for-all and distort the price structure. Painful adjustments of price differentials were necessary in late 1974 and may prove inevitable this year because producers on both occasions entered the slack market with the grossly distorted price structure.

A slack market need not be particularly troublesome for OPEC once the structure of price differentials is adjusted. In 1974–75 the adjustments took place over six to nine months and thereafter OPEC survived a long period of relative stagnation in demand. Between 1975 and late 1978, for almost four successive years the oil market was fairly slack. As soon as the price structure found its equilibrium the market began to distribute in proportion among the various producers any reduction in aggregate demand. There were no strains within OPEC during this long slack period.

We are now in a position to draw a number of lessons on pricing issues.

The first proposition is that OPEC can fulfil many of its objectives through the administration of the reference price of the marker crude. As mentioned earlier the determination of the price level of the marker is not dependent on the vagaries of the supply-demand balance. Gradual increases in the price of the marker throughout phases of shortages and gluts are both possible and desirable. They would satisfy the first pricing objective of OPEC in protecting revenues from the bite of inflation and the fluctuations of exchange rate. If applied in a gradual manner they would not disrupt the world economy. Further, graduated rises will signal to consumers that oil may become scarcer in the long run and will correct the misleading impressions of glut which may be created in the short term by temporary recessions. OPEC so far has not followed systematically such a policy. The marker price was raised significantly in 1973–74 and in 1979–80, but it was allowed to stagnate in the intervening period save for two or three very minor and nominal increases. It is possible to argue with the benefit of hindsight that this passive behaviour displayed throughout 1975–78 was not well justified. It was motivated by a concern for the world economy. Yet the decline in the real price of oil which characterized this period does not seem to have induced a recovery nor did it have a visible effect on the rate of inflation. The decline in the price of oil failed both producers and consumers. The former suffered a drop in the real value of their earnings and the latter were not provided with much needed incentives for oil conservation and investment in alternative sources of energy.

The OPEC Committee on long-term strategies has considered these problems and proposed a formula which provides for small but regular increases in the price of oil. The pricing policies of OPEC in the 1980s will be substantially different than in the past years if the recommendations of the Committee are approved by the Summit and thereafter implemented consistently. A successful performance on this score would mark a considerable improvement on the past as it would enable OPEC to fulfil in better ways its two important pricing objectives to the advantage of its members and the world community.

The second lesson to be drawn from past experience relates to the problem of price differentials. OPEC has been long concerned with this issue and much money was spent on econometric models in a vain search for a recipe. We must have learnt by now that models, however sophisticated, cannot provide feasible formulae for the automatic administration of price differentials. The market must be allowed to play a role in inducing small marginal adjustments to a fairly orderly and flexible structure of price differentials. No serious strains are involved when the initial price structure reflects correctly differences in relative values and the adjustments are normal responses to small changes in certain parameters such as fluctuations in tanker freight rates or seasonal variations in the demand for heavy and light products. The problem of differentials arises when the price structure is grossly distorted, and we have argued that significant distortions affect the system when the market is tight. As soon as the demand cycle takes a downward turn market forces begin to exercise a powerful influence on the system to remove these distortions. Serious strains develop because the large adjustments required cannot be achieved without disruptive shifts in the distribution of supplies. Those producers who happen to be overpriced risk losing suddenly a big proportion of their market shares and their price responses to this loss can induce competitive underbidding. The only way to avoid such a situation is to prevent the emergence of a distorted price structure in a tight market. This is never easy and may not always be possible. A more realistic aim is to try to minimize the extent of price distortions during a shortage and to moderate the harsh impact of market forces on the system in the subsequent period when excess demand begins to set in. These aims can only be achieved through a collective production policy.

A production policy designed to cope alternatively with shortages and gluts involves two components. It must be able first to cope with excess demand, the original cause of price distortions. This means that OPEC countries must always carry a certain amount of reserve capacity and have a plan specifying the ways and the conditions under which they will draw on this capacity to meet partially at least a critical imbalance in demand. The purpose of such a plan is not to mop out continually any excess demand that happens to emerge nor to prevent prices from rising, but to attentuate the impact of serious crises.

The second component of the production policy would aim at smoothing the transition from shortages to glut. The policy could be designed in a very simple way with *pro-rata* reductions in supplies by all members in response to a decline in demand. The *pro-rata* reduction would be applied to the most immediate production peak. This policy need not prevent market forces from inducing an adjustment of price differentials, but would stretch the period through which those adjustments take place and minimize the strains caused by violent shifts in the distribution of supplies of certain member countries. In the past few countries were able to absorb the bulk of such reduction and thus were playing the role of swinging suppliers. As those countries' production approaches their optimum capacity any reduction in offtake should in future be shared pro-rata by all member countries.

To conclude, it is possible in theory to choose between price administration and output programming. In the real world price administration must be underpinned in some way, by a production policy. Such a policy is needed not only for the problem of price differentials but for a successful implementation of the formulae proposed by the strategy committee for the price of the marker crude.

*Market Structures*

I now propose to examine a second issue, that of marketing relationships between OPEC producers and buyers of crude oil. In discussing prices, I suggested policies for the future which in some important respects are different from those followed since 1973. On Marketing I am inclined to favour a continuation of recent developments rather than substantial changes.

The price revolution of 1973 was followed by a significant transformation in the relationship between Governments and companies. I shall not dwell for very long on this subject which has been discussed in detail last week by distinguished speakers. Let me simply highlight a few important points.

The recent history of the world oil industry can be interpreted as a process during which producing states reclaimed sovereign rights over oil which they had alienated under the initial concession agreements. These rights are not limited to autonomy in pricing decisions. They involve the whole management of the crude oil industry from exploration and development to production and marketing. Governments began to exercise the right to define allowable levels of output in the early 1970s. In 1973 sovereignty over prices was dramatically reclaimed. The emergence of state agencies as direct marketeers of crude oil did not follow a uniform pattern throughout OPEC. This development began early on in some countries, much later in others. There are still today considerable variations in the extent of marketing to non-concessionnaires as between the different OPEC countries. There is no doubt, however, that the role of governments in this field began to expand after 1975 and that it received a very significant boost in 1979 as a result of the Iranian crisis.

A typical (though not universal) situation today is one in which an OPEC country sells its oil to 20 or 30 different buyers, some major oil companies and some independent or NOCs. Government to Government deals have also become common. This new situation is advantageous to producers in a number of respects.

The first and major advantage is that it avoids the dangers of monopsony. It is preferable to face some competition between buyers than to depend on one or two. A second major advantage is that diversification provides access to different sources of information about the behaviour of the oil market and the situation of the world oil industry. Information is a valuable commodity and the dependence of OPEC countries in the days of the concession system on a single source (the Majors) had obvious drawbacks. A third aspect is an enhanced ability to operate in different markets and with different types of companies and to benefit from the diversity of opportunities.

All in all these changes in marketing relationships are a healthy and beneficial development. Contrary to certain prejudiced views, it does not involve the total demise of the Majors nor the transformation of the crude market into a rigid set of official bilateral deals. The share of the major companies in offtake from OPEC countries of course has declined but they continue to enjoy an important place. To dispense with them is to nobody's advantage as they have much to offer in terms of technology, logistics and experience. Other companies provide some of these services as well as the benefits of diversification. They can all contribute in different ways and it seems sensible to retain some access to all of them.

It is also wrong to suggest that the new marketing structure has become so rigid as to constrain adjustments and to foster inefficiencies. Government to Government deals are not more rigid than contracts between a producer state and an oil company. Most countries utilize the same contractual format irrespective of the customer. Variation and phasing out clauses are generally identical for all. And it is only through a bizarre twist of economic logic that one can suggest greater inefficiencies in diversification and buyers' competition than in monopsony!

It is true that the increase in the number of buyers has raised the average requirement for working stocks in the world and the tanker capacity needed for the transport of a given volume of crude. But these are once-for-all changes and I have little doubt that the industry will settle in the new system without further ado.

My perception for the future of crude oil marketing is a continuation of the present trends in countries which have not yet reached their preferred pattern of diversification. Once achieved this new pattern will tend to remain stable.

Progress by OPEC countries remains to be made in two important areas. The National Oil Companies have not yet fully developed into efficient and mature organization. They face all the familiar problems encountered in developing countries: shortages of managerial and technical skills and inadequate back-up from the country's institutional and physical infrastructure. In most countries – including by the way the United Kingdom – the problem of the role of NOCs and of their relation with the Government Departments is not clearly solved.

The second issue is that of development downstream. Some OPEC countries are investing considerable sums in refineries, petrochemical plants and tankers. Timid attempts are made to invest in the oil industry abroad. I feel, however, that strategies linking downstream investment and activities with the marketing of crude oil have not been yet worked out. It is not clear whether crude should be dealt with separately from its products or whether an integrated approach involves advantages. The whole issue of downstream oil is of considerable importance because the industrialization of OPEC countries is bound to depend to a large extent on the processing of their major raw materials.

## The International Relations of OPEC

The price revolution of 1973 has transformed the international image of OPEC and introduced oil-producing countries as powerful actors on the world scene. A new pattern of relationship between OPEC members and the rest of the world has emerged. This situation involves both opportunities and responsibilities.

It is useful to distinguish in this context the relations of OPEC with industrialized and with developing countries. The industrialized countries are fundamentally antagonistic towards OPEC because they resent their dependence on imports of oil on countries which have succeeded in administering its price. A few in the industrialized world have come to recognize that OPEC performs, though imperfectly, a useful role in the management of the energy crisis; but they are a minority. The developing countries are far the most ambivalent. High prices of oil impose a heavy burden on their economies. But OPEC success in reversing the behaviour of the terms of trade in favour of a primary producer strikes a very sympathetic and profound chord.

Little attempt was made at defining objectives and at working out the implications of OPEC international relationship until the formation of the Long Term Strategy Committee. OPEC after all assumed this responsibility only a few years ago. The 1980s will probably see a significant change of behaviour in this respect. The work of the Committee reflects and in turn crystallises perceptions within OPEC about opportunities for new patterns of relationships with these two groups of countries and about the need to initiate actions which could involve a broad range of mutual benefits.

OPEC countries have potentially considerable leverage with the industrialized world because of their oil power. This leverage could be used with discretion to improve the terms on which OPEC countries acquire (a) the technology, the goods and services necessary for their economic development and (b) financial assets exchanged against their surplus revenues.

Economic development is the main objective of OPEC countries and will be as vital to them in the long run as oil is to the industrialized world in the medium term. Economic development is the long term substitute for the producers' oil wealth; and oil is the current substitute of the energy sources which will fuel in the future the industry of advanced countries. It is essential that OPEC obtains today the inputs of economic development on advantageous terms and the only opportunity to improve the condition of exchange in favour of OPEC arises now when it is vested with oil power. In my opinion, the issues related to economic development shall in the future be the focal point for OPEC and not prices in monetary terms.

A similar argument can be made for financial assets. These placements which earn a negative real rate of interest and whose value, therefore, slowly evaporates with the passage of time are assets destined to replace in the national portfolio of wasting oil. It is puzzling that OPEC countries have accepted for so long such unfavourable terms for their monetary investments in the West.

Finally, leverage can be applied to shake the industrialized countries out of their growing indifference towards the Third World. OPEC has provided such aid since 1973 though much remains to be done to improve quality and distribution of the assistance. OPEC intends to do more and to do it better in the years to come. But there is a real danger that additional help to developing countries does not increase by its full amount the resources available to them. The industrialized nations are inclined to think that OPEC aid excuses them from the effort and is meant to relieve them from the obligation to pay. This attitude inhibits to some extent OPEC aid donors unwilling to see their additional contribution diverted away from the Third World.

There is significant scope for an active role for OPEC within the Third World. Several schemes are being presently debated which would extend the means and forms of financial assistance. A question is often asked: what is the objective of OPEC financial solidarity with other developing countries? To my mind the answer is fairly simple and lies in the mutuality of both political and economic interests. In fact they are one group. OPEC countries value the political support of other developing nations and in turn add to the political weight of the Third World in international fora. The prospects of economic development in OPEC countries are enhanced as other developing countries begin to prosper thus providing great opportunities for investment and trade and improved supplies of manpower and expertise.

The time has come to conclude. OPEC has new, important and challenging tasks to perform in the future. Much has been learnt from the experience of the past seven or eight years. The timely work of the Long Term Strategy Committee has defined with imagination and competence objectives and proposed methods. Of course, it is easier to chart as I have done today the course that ought to be followed than to implement successfully. There are constraints on OPEC countries which restrict them in the exercise of functions perceived as essential. One major constraint is economic underdevelopment and a second is foreign political interference. It is possible that OPEC will find itself unable in the 1980s to perform in a significantly different way its international role as price administrator, agent of structural changes and catalyst of fruitful relationships with poor and industrialized nations. This would be, of course, OPEC's loss but I dare say that the rest of the world would be the lesser for the failure.

# CHAPTER 11

# THE CHALLENGE OF THE OIL MARKET

*The following is the substantial text of an address delivered at the Third Oxford Energy Seminar, 24 September 1981.*

It gives me great pleasure to address this third session of the Oxford Energy Seminar. I have been associated with the Seminar since its foundation. I witnessed its birth four years ago in my former capacity as Secretary General of OPEC, and I had the privilege to speak every year to the distinguished audiences which assemble annually at St. Catherines for fruitful debates. This privilege, however, carries certain risks. A regular speaker is in danger of repeating himself year after year. Fortunately this risk of repetition can be avoided to some extent because the circumstances of the oil market and the nature of the problems facing the producing countries seem to change continually. When the First Oxford Seminar gathered in 1979 the oil market was tight, petroleum prices were rising and oil-consuming countries and oil companies were seeking contracts from producing countries at almost any price. There was concern about the security of oil supplies leading to an unprecedented build-up of inventories. Government to Government deals became fashionable and many a Government or a company was prepared to offer technical assistance and special trade arrangement or to participate in joint ventures in OPEC countries in order to gain access to oil supplies.

The Second Oxford Seminar took place in 1980 in a different environment. There was less concern about supplies, because inventories had reached a historical peak and because the spot price of crude oil had begun to decline. As the short-term situation was easing many of us began to take a more relaxed view about immediate crises and to give some attention to the long-term. OPEC at that time, was preparing to celebrate its 20th anniversary and to mark the occasion with the adoption of an ambitious and far reaching long-term strategy.

The Third Seminar, just one year later, is taking place in a different environment. The oil market is slack. The demand for OPEC oil has fallen dramatically this summer to a very low level. Oil companies, rather than seeking new contracts with oil-producing countries, are walking out on previous agreements, rather than acquiring oil they prefer to run down their inventories. Spot prices have come down to the level of the marker crude; and the OPEC oil price structure is subjected to strong market pressures. The OPEC long term strategy is eliciting much less public interest and there seems to be less concern with such issues as the security of and the access to oil supplies.

These changes in circumstances enable us to select every year a different theme. The slackness of the oil market is a topical issue. The oil-producing countries are faced with a challenge, and it will be interesting to examine how they can successfully respond to it.

So far I have emphasized the apparent changes in the oil situation which we all observe from year to year. These changes however do not preclude the survival of certain similarities. The current situation for example resembles in many respects other recent episodes of the turbulent oil scene. Further strong elements of continuity underly all these changes. The oil market continues to function whether the supply-demand balance is slack or tight. OPEC remains a strong institution with a long experience of crises and successes. The oil-producing countries face, today as they did yesterday, the same long term problems of economic development and of excessive reliance on a depletable resource. The oil-consuming countries on one day seem concerned about the security of oil supplies and on another totally forgetful of the significance of the issue. Yet, the threat of a long-term energy crunch never fades away. I propose to treat my subject in two parts. The first part involves a comparison of the current market situation with a similar episode of the recent past. The second part analyses some of the challenges which face OPEC in the short and the long run.

Two episodes of recent oil history have shown common features. The first episode covers the period 1973–75 and the second began in late 1978 and continues to date. In both cases a tightening of the oil market led very rapidly to a pre-explosion which was then followed after a short while by a significant drop in world oil demand.

Let us recount briefly and compare the two episodes. At the start of the chain of events, two developments unfavourable to the economic interests of the oil-producing countries were taking place. The exchange value of the dollar in relation to other currencies was falling and the rate of world inflation was rising rapidly fuelled by an economic boom. Oil-producing countries were becoming deeply dissatisfied with the erosion in the real value of the unit price of oil. But they were only able to alter the situation in their favour when they enjoyed market powers. This power arises when the oil supply-demand balance tightens on the market, and this is precisely what happened in late 1972 and throughout 1973. The market began to tighten well before the Arab-Israeli war. This is an important fact which many commentators on oil affairs either seem unaware of, or prefer to ignore. The tightening of the market was being signalled by all the usual indicators: tanker freight rates, product prices and spot crude prices. To give one example, the world average realized price of crude had risen by $2 a barrel between October 1972 and September 1973, but to be compared with an increase of less than 50 cents in the posted price of Arabian Light. The OPEC price rise of October 1973 is a natural result of these developments. It is vain to seek a political explanation for this price rise which was entirely due to the tightening of the market. The Arab-Israeli War and the production cut-back tightened the market further. They created much uncertainty among oil buyers who began to bid-up spot prices to unprecedented levels. Transactions at $17 per barrel and above were reported. OPEC followed suit by raising the posted price of Arabian Light to $11.65 in the same direction as the spot, but well below the maximum level attained.

The production cut-backs imposed by Arab oil producers in October 1973 were relaxed by the end of December and terminated in March 1974. Since the squeeze on oil supplies was relatively short-lived there was no price escalation after the price rises of December 1973 apart from a few minor adjustments, aimed mainly to reduce the gap between the cost of crudes to concessionaires and direct sales to third parties. This had resulted in what is termed today as Government Selling Prices (G.S.P.).

At the beginning of 1974, as the Arab measures were being lifted, world demand for oil rose. But the increase in demand which took place during the first half of 1974 was largely due to the building-up of inventories. The irony of oil history is that the level of inventories held by companies and governments of consuming countries always seem inadequate before a crisis, and that they tend to be built up to very high levels as soon as the crisis is over. These inventories are then run down after a fairly short lapse of time – say, six months or a year – because the next crisis fails to materialize and because the memories of the first commotion seems to fade away very quickly.

During the period from October 1973 to mid-1974, when the market was tight, because of the supply disruption and the increase in inventory demand, the OPEC price structure became distorted. Oil producers with high quality oil such as Libya or Abu Dhabi were able to charge prices involving very high differentials above the marker crude. The reason is simple. A tight market is a free for all. Every producer can easily sell any available amount of oil to eager buyers willing to bid up to high levels. The concept of normal differentials is meaningless in a tight market. Normal differentials can only be obtained when buyers have a choice at the margin between different varieties of crude, that is when the market is either finely balanced or slack.

When demand exceeds supply there is no room for choice. Everbody is willing to grab any amount of any variety of crude that is available. Oil producers can determine their own prices independently from each other when they find themselves in a sellers market. Those who happen to be tough bargainers manage to get higher prices than others irrespective of quality differentials. This is precisely what happened in late 1973, in the first half of 1974. During this period, the posted price of Libyan 40°, for example, was $3 above that of Arabian Light, while the "normal" differential for Libyan crude at the time was generally estimated at 80 cents; Abu Dhabi's price for Murban involved a $1 mark-up against a "normal" differential of 40 cents.

Neither Libya, nor Abu Dhabi, suffered a reduction in offtake until mid-1974 when the market was tight. But this situation changed suddenly in the second half of 1974.

The price rises of late 1973, as mentioned earlier, were followed immediately by an increase in demand for inventory purposes, but this proved to be a temporary phenomenon. There was a drop in world oil consumption which led, from mid-1974 to the beginning of 1976, to a fall in oil demand. The reduction in demand was largely due to an economic recession. There was no immediate impact on oil from other sources of energy as their development involves long lead-in times. The early effects of conservation were also minimal.

The features of the oil market in the second half of 1974 are of particular interest to our story because they shed considerable light on what is happening today. There was decline in oil demand which caused a significant slackening of the market. The OPEC price structure in mid-1974 when the market ceased to be tight was extremely distorted. Oil companies immediately began to reduce their offtake from high-price producers while continuing to lift as usual from the cheaper sources. These reductions in offtake at times were very substantial, of the order of 50 or 60 per cent. The producers who suffered those losses in export volumes responded by lowering their prices in steps over a period of six or seven months until they reached the correct level of relation to the market. These adjustments were done by individual producers independently. There was no serious attempt to agree collectively within OPEC on a revision of the price structure. Most remarkably the price adjustments restored very quickly export shares. An equilibrium, though at lower level, was attained during 1975 which lasted without trouble or disruption until the Iranian crises of 1979.

The following data give an idea of the magnitude of these developments. OPEC production in the six months July–December 1974 was 10 per cent below the September 1973 level. In December 1974, the aggregate drop from September 1973, taken as a base, was 13.5 per cent. But this reduction was not equally distributed between producers. Average production in Saudi Arabia and Iran during July–December 1974 was slightly higher than in September 1973. But Libya suffered a 46 per cent reduction in output, Venezuela 17 per cent and Qatar 13 per cent. In December 1974, the reduction was considerably greater: Libya 56 per cent, Algeria 18 per cent, Venezuela 20 per cent and Indonesia 22 per cent. In January 1975, Abu Dhabi suffered in a single month a sudden drop of 37 per cent in its production.

The price adjustments made by these producers were very substantial. Libya lowered its prices in steps from a posting of $15.76 in early 1974 to a G.S.P. $12.30 in mid-1975. This adjustment was equivalent to a 22 per cent reduction in revenue per barrel of oil. Abu Dhabi lowered its price by $.95 in January 1975 from a posting of $12.63 a year earlier to a G.S.P. of $10.87, equivalent to a 10 per cent reduction in unit revenue. In percentage terms, the changes required today to adjust the OPEC price structure are much smaller; yet they are eliciting greater resistance from member countries.

The 1973–76 episode which we have been analysing ends with a significant increase in world oil demand in 1976. Oil demand in the world (excluding communist countries) dropped from 47.9 MMb/d in 1973 to 46.3 MMb/d in 1974 and 45.2 MMb/d in 1975. But this "demand crisis" ended after 18 months. In 1976, demand rose to 48 MMb/d, reaching a higher level than the peak of 1973. Now compare this episode with developments in the world oil market from 1978 to date. You will immediately recognize striking similarities. Many features of the two periods are virtually identical. The same patterns of behaviour seem to recur, the same patterns of responses seem to appear.

In 1978, oil-producing countries were dissatisfied with world inflation and another weakening of the dollar. The market had been fairly balanced if not a bit slack since the beginning of 1977. However, by mid-1978, before the Iranian crisis, certain timid signs of tightening started to emerge. Product prices began to rise and spot prices of crude oil to firm up. The Iranian crisis provoked a price explosion, largely due to sudden structural changes in the industry and to oil buyers' reactions to uncertainty. The supply disruption due to the Iranian crisis was in fact very short because other oil-producing countries increased their production in compensation for the Iranian shortfall. As in 1974, oil companies were induced to build-up their inventories immediately after the disruption. The build-up started very early on, in the second quarter of 1979 and continued until the third quarter of 1980. This was a major build-up, taking inventories to a historically unprecedented high level. Thus demand for oil increased for inventory purposes at a time when available supplies were more than sufficient to meet consumption requirement. This increased demand fuelled further price rises.

As in 1973, spot prices led the movement throughout 1979. OPEC followed suit with a lag, and official price rises fell short of the levels attained on the spot market. The convergence between spot and official prices occurred much later when the market slackened and spot prices sharply declined while OPEC was still in the process of catching up.

As in 1974, the OPEC price structure became distorted for exactly the same reasons. In a tight situation every producer is in a sellers' market and has considerable freedom in matters of pricing.

The oil price rises were followed by a decline in world consumption, initially concealed by the rise in inventory demand. As soon as the build-up ended the fall in oil consumption manifested itself as a downward demand trend. In this respect, the situation today resembles that of 1975. However, the factors affecting demand have different weights than in 1975. The effects of conservation and displacement of OPEC oil by a substitute appears to be larger. As in the second half of 1974, oil companies are walking out on high price producers. During this summer export volumes fell significantly in Nigeria, Lybia, Mexico, Kuwait and elsewhere. By contrast, exports remained unaffected in countries where prices are correctly defined in relation to the marker. These include Saudi Arabia, the North Sea producers and some others.

The decrease in world oil demand has not been distributed in a proportional manner between the various oil-producing countries. A few producers are carrying the whole brunt of the demand shock while others are escaping unscathed. As you all know, this is in part consequence of the distortions in the price structure, which are themselves a consequence of pricing policies during the phase of market tightness which preceded the current slack.

When the market is slack and the price structure distorted the fall in world oil demand is larger than the decrease in world oil consumption. Oil companies rather than purchasing oil they prefer, if the need arises, to draw down their inventories. My view is that the drop in demand this summer was much larger than the average drop in consumption measured on an annual basis. This means that oil companies cannot maintain for very long their present stance. They would not be able to satisfy the requirements of world oil consumption during the coming winter if their liftings from OPEC countries and from such countries as Mexico were to be maintained at the levels which were obtained this July. Their options are either to draw inventories down or to raise their nominations for offtake from some of the high-price producers. If the offtake volumes of July were to be maintained in the winter, inventories would have to be drawn down 3.5 to 4.0 MMb/d.

You can readily recognise that the inventory cycle of 1978–81 has the same qualitative features as the inventory cycle of 1973–75. The supply crisis catches everybody napping with inadequate cushion from inventories. For a short period during the supply disruption inventories are naturally run down.

As soon as the worst impact of the supply disruption is absorbed everybody attempts to build up stocks to very high levels; and the process prolongs the period during which the market is tight and fuels further price increases. The crisis of 1973–1974 and 1979–1980 each involved two stages: a supply disruption followed by a rise in demand entirely attributable to a frantic urge to build up stocks. But stock levels cannot be increased for ever. Sooner or later the process has to stop and the demand for oil begins to decline for the simple reason that inventory demand dwindles down to zero. This fall in demand, which is usually compounded by a decrease in consumption due to the price rise, produces a glut on the market. The glut enables companies to shop around. They start fighting against high price producers and use their inventory cushion as a weapon in the price war. They draw down their inventories, and this process results in a further reduction in oil demand which aggravates the glut. Finally, the glut leads the industry to expect a fall in the real price of oil; and the expectation of falling prices leads to a further reduction in stocks. All that happened in 1974–75 as described here and we are witnessing today once again the same phenomenon.

The historical comparison enables us to assess the nature of the challenge which faces OPEC in the oil market. The ability to cope with a slack market is significantly enhanced by an orderly price structure. This was demonstrated with great clarity in 1975. The oil market stabilized as soon as the various oil-producing countries adjusted their differentials from the marker crude to the correct level. An orderly price structure has two main beneficial effects on the market. First, the reduction in world oil demand would be distributed in equitable proportion between various producers. A situation in which every producer suffers the same relative reduction in offtake as everybody else is infinitely more tolerable than the current situation where some countries have to cope with a very large drop in export volumes while others escape unscathed.

Secondly, the reduction in oil demand would not be as large with an orderly price structure because companies would not need to use the inventory weapon to force price adjustments.

Our message is very clear and is perfectly understood by all concerned. The most urgent task facing OPEC is to sort out the price structure. As you know this can be achieved in a variety of ways: *either* by retaining the current price of the marker crude and lowering the differentials, *or* by raising the price of the marker crude, *or* by some agreed combination of both measures. The last solution offers the best chance of success as it involves matching concessions from all sides.

So far, the compromise has eluded OPEC because every oil producing country feels committed for both political and economic reasons to its own G.S.P. Non-OPEC countries, such as Mexico, share the same attitude and find price adjustments extremely painful. It is unfortunate that the notion of flexible differentials relating to a fixed marker price which was generally accepted during the 1973–76 episode should have given way to rigid views about G.S.P.

Flexible differentials allow every producing country to take advantage of a tight market as best it can, and to adjust rapidly and effectively to the conditions of a slack market without endangering OPEC solidarity and without jeopardizing OPEC hard won achievements.

Attempts at establishing an orderly price structure during an oil shortage are vain, because they lead to artificial and meaningless arrangements which complicate subsequently the vital task of adjusting the price structure during the slack. No such attempts were made in 1973–74. The marker price was defined by OPEC and the differentials though formally approved by OPEC were in practise freely determined by the country concerned. Nobody seriously tried to conceal this freedom by introducing concepts such as a ceiling on differentials (which was unenforceable and which was repeatedly ignored), or a dual marker price (which in a thinly disguised way means unequal differentials for similar varieties of crude). In 1979–80 all these notions were used to show in a very artificial way that the price structure obeyed a certain order. The pretence served no useful purpose at the time; and it positively complicated OPEC's task in restoring the price structure when the market later became slack.

My belief is that OPEC would easily be able to cope with the present challenge of the oil market as soon as an agreement on the price structure is reached. I also believe that such an agreement will emerge in the near future on the basis of a compromise involving a rise in the marker price and adjustment of differentials. The strong incentives to reach such an agreement arise a) from the commitment of all member countries to OPEC and b) from an understanding that market pressures are better resisted with the protection of an orderly price structure.

OPEC's recent history reinforces my optimism. The organisation was able to cope very successfully with adverse market conditions in late 1974 and 1975. OPEC held the price line, by which I mean the reference prices of crude, despite a significant fall in world demand and the initial handicap arising from a disorderly price structure. The price adjustments required at that time were much more significant than those which are called for today. OPEC held the price line in 1977–78 when market conditions were indifferent. The difficulties we are facing today on adjusting the price structure are largely of our own making. We know that our past achievements were made possible through solidarity; and that these achievements can be threatened if a consensus fails to emerge. A genuine consensus is only reached when all sides feel that they have contributed to it through mutual concessions justified by an over-riding common interest.

In conclusion, I shall make some comments about the challenge of the oil market in the long run. It is unwise to forecast developments of the oil market in the years ahead. Some observers believe that the market will remain slack for a long time. Others think that the oil surplus will disappear sooner or later and that a tight market may re-emerge after a year or two.

A reduction in world oil demand may turn out to be a blessing in disguise. Our countries are not really benefiting from high levels of oil production which deplete natural reserves for the sake of financial accumulation. Our economic development is a slow and long affair, and the surplus revenues which accrue as a result of high depletion are causing more harm than good to the development prospects.

Whether the market in the future is tight or slack we should begin to consider seriously how best to use our resources for the building up of our economies. To sell oil only for cash is not very satisfactory. Access to our oil supplies should be related to effective contribution made to our economic development through investments in joint ventures, training and transfer of technology, especially in the hydrocarbon sector. In 1979–80 when the oil market was tight oil companies and Governments of consuming countries began to show an interest in such ideas. This interest seems to have disappeared as soon as the market became slack. Yet economic development requires sustained efforts and cannot be made to depend on such volatile intentions.

If OPEC countries want to get value for money they should adopt marketing policies which ensure preferential access to those who are willing to contribute to their development. We used to give in the past preferential access to some companies over others for not very good reasons. We should perhaps revert to this policy not to favour the concessionaires of the past but those who want to co-operate with us for the achievement of vital objectives.

In short the main challenge facing us in the long run is not so much how to play the market in its ups and downs. We should be able to manage that, and in fact we did in the past. It is rather how to transform our oil resources into the real asset of economic development.

# CHAPTER 12

# PERSPECTIVES ON THE OIL MARKET

*The following is the substantial text of a lecture delivered to the Fourth Oxford Energy Seminar, 1 September 1982.*

It gives me pleasure to participate and to contribute to the programme of the Fourth Oxford Energy Seminar. Every year since the establishment of the Seminar I have been given the opportunity to present my views on the role of OPEC on the challenges and opportunities which face this organisation of oil producing countries, and on the relationships between OPEC and the world petroleum market.

To my surprise I am finding that every year there is something new to say. The reason is that the world energy situation is in a constant state of flux. The conditions which describe the market, by which I mean the demand for and the supply of oil, the structure of the market and the behaviour of buyers and sellers, seems to be changing continually and in a significant manner.

Not long ago, in 1979 and 1980, the world petroleum market was characterized by tight supplies. The Iranian Revolution induced serious fears about the security of oil supplies. There was a rush for oil from companies and Governments alike. Inventories were built up initially for precautionary motives, and later for speculative reasons. Governments sought to obtain oil directly from oil-producing countries through privileged deals. In the process the price of oil rose manifold. Some observers have referred to this episode as the second oil price revolution.

In 1981 the situation became radically different. The oil market was slack because of important changes in demand conditions. These changes may be described as follows:

*First*, there was a decline in final consumption of oil products. This decline may have begun in 1979 but did not attract much attention at the time because everybody was concerned by the supply disruption in Iran and later by the Iran – Iraq war at its inception.

*Secondly*, there was a reversal in the pattern of inventory demand. As stocks had reached an historic high level, companies ceased to build up their inventories and began to look for ways to reduce the volume of their holdings. As you know, a change from a stock build up to a stock drawdown has a compound effect on total demand.

The severity of the demand decline in 1981 is better understood if we recall that it succeeded in depressing the market even though supplies were being curtailed by the destructive effects of the Iraq-Iran war. OPEC did not respond very well to the conditions of the oil market which prevailed in 1981. The oil-producing countries entered the period of slack demand with a distorted price structure which had developed in 1979–80 when the oil market was tight. By "distorted price structure" I mean that the price differentials for various types of crude oil – light, medium, heavy etc. – were out of line relatively to the marker crude and to one another. These distortions came about in 1979–80 when every producer felt free to charge any price he wished. The excuse is that demand was buoyant, that the customers were anxious to secure supplies and that everybody was bidding up prices with ot without reason. When buyers lose their heads, sellers cannot be expected to hold back and show restraint. We are all human after all and the temptation to join the heady dance in a free-for-all is almost irresistible.

Distortions in the price structure do not matter when the market is tight, since everybody then is able to sell as much as he wants or as much as he can. Good luck to those who get an additional dollar or two per barrel; and no tears for those who sell at slightly lower prices because they are doing very well anyway.

But a distorted price structure can have adverse consequences for OPEC member countries in a slack oil market. When demand is depressed buyers have the upper hand. They can shop around because every seller suffers from excess capacity and can meet additional demand for his oil. Buyers will naturally look for oil sold at the most advantageous price. They will withdraw their custom from countries which charge an inflated differential and shift their purchases in favour of other countries. The result is that the drop in total demand is unevenly distributed between the various producing countries. Some find themselves selling as much as before while others suffer big reductions in the volume of their export sales. This uneven distribution of demand lead to a reduction among OPEC countries of their ability to administer effectively the price of oil in international trade.

This is precisely what happened in 1981. The price structure was distorted, and though the market was slack many OPEC countries refused for many months to adjust their price differentials and to help bring-about an orderly price structure.

In my lecture last year to the Third Oxford Energy Seminar I identified this problem and emphasised the need for a solution. To quote from last years paper:

"Our message is very clear and is perfectly understood by all concerned. The most urgent task facing OPEC is to sort out the price structure."

I added:

"So far the compromise has eluded OPEC because every oil-producing country feels committed for both political and economic reason to its own G.S.P. Non-OPEC countries such as Mexico share the same attitude and find price adjustments extremely painful. It is unfortunate that the notion of flexible differentials relating to a fixed market price which was generally accepted in earlier years should have given way to rigid views about G.S.P."

We were then in September 1981. A few weeks later, in October, OPEC reached an agreement about the price structure and final adjustments were made in the Abu Dhabi meeting in December 1981. Meanwhile much damage had been done. Saudi Arabia which was seeking all along – indeed since the beginning of 1979 – an orderly price behaviour attempted to influence the behaviour of other members by producing oil at maximum capacity. This production policy was designed to prompt a uniformity of the price structure. As the official price of Arabian light was lower during that period than the deemed market price, liftings from Saudi Arabia naturally increased near to maximum capacity. But there were side effects. The additional oil found its way to inventories. All that could have been avoided if OPEC members had fully appraised the importance of an orderly price structure. It is wrong to believe that every G.S.P. is a sacred political symbol. OPEC's role is to define and to hold a fixed reference price – that of the marker crude. All other prices should relate to this reference in a way which correctly reflects the relative market valuation of the quality differential.

As these relative values tend to vary from time to time it is essential to introduce a certain flexibility in the price structure. But it is also essential that this flexibility be used to procure the correct adjustments especially when the market is slacker. Those who insist on a price in excess of the correct differential cause damage both to their own interests and to the interests of OPEC as a whole. And those who underprice their oil relatively to the correct differential are in fact competing against their colleagues and undermining the solidarity of OPEC.

The market conditions in 1982 turned out to be even worse than those prevailing in 1981. Having just succeeded in adjusting the oil price structure OPEC found itself facing a major demand crisis. The current situation may be analysed as follows: World demand for oil seems to be on a downhill slide. It is important to distinguish two components in demand, one relating to consumption and one to stocks. Let us first talk about consumption.

There is no doubt that world consumption of oil is falling. In my judgement this is largely due to economic recession, though conservation and substitution of coal and gas for oil are also playing a part. Contrary to the conventional wisdom prevailing in oil-consuming countries OPEC does not have much influence on the demand for oil and OPEC countries were not particularly happy with the rapid rise in oil demand which occurred in the 1960s and until 1973. They feared the long-term economic consequence of an early depletion of their reserves. They resented the open-ended commitment of having to satisfy the ever-increasing requirements of an oil greedy world at the risk of jeopardising their economic development and the interests of their future generations. The price rises of 1973 were not an act of greed but the inevitable economic consequence of expanding demand drying up available supplies.

OPEC countries are not happy either with a demand situation which reduces their output to some 55 per cent of their 1979 output. Some say that the price of oil is responsible for this situation. We all know however that the price of oil is but one factor among many which influence the state of the world economy and in turn determine the demand for petroleum. The other factors are the policies pursued by many Governments to combat inflationary tendencies which began to develop well before the oil price revolution. Low productivity growth, high interest rates, lack of confidence among investors, workers' unwillingness to compromise about wage increases, are among the many ills of the world economy. OPEC cannot cure these ills.

To suggest that OPEC should do its bit unilaterally and lower the price of oil is disingenuous. First, nobody can be certain that such a reduction will revive the world economy and most likely it will not. Secondly, too many Governments of consumer countries are ready to neutralise the effects of such a reduction by increasing their domestic taxes or by imposing a countervailing import tariff on oil. The growth of economic literature on those tax measures particularly in the United States provides ample evidence about the current interest in consuming countries for these adversary strategies.

I fear therefore that OPEC, *acting on its own*, can do very little about oil demand. We would of course avoid policies which may worsen the outlook. For the rest we have no other option but to sit it through. In reality maintaining prices at their current monetary terms amounts to defacto reduction of prices in real terms.

Let me now add a word about demand for stocks. In 1979 and 1980 the oil markets became overheated, not because consumption requirements exceeded available supplies but because stock building took place at an unprecedented rate. It is fashionable to blame OPEC for the price rises of 1979–80. But if one takes a dispassionate look at the matter it would readily appear that the price rise was demand – led by the producing countries – and I include among those the United Kingdom, Mexico and non-OPEC producers – and OPEC did not engineer the rise. They followed demand signals in the market. Ironically, the non-OPEC producers, especially the UK, proved themselves to be quicker in their responses and less inhibited than OPEC countries in raising prices. Moreover, this surge in demand was entirely due to the desire to replenish stocks unwisely depleted in 1978.

Stocks also played a major role in 1982. There was a massive drawdown which aggravated the demand situation already depressed by the decline in consumption. Companies have strong incentives to get rid of stocks. First, interest rates are high which means that stocks are significant. Secondly, companies are having cash flow problems which can be eased to some extent by liquidating stocks. Thirdly, they expect prices to fall at least in real terms. Fourthly, as consumption is declining the amount of stocks required to cover a given number of days of consumption is also diminishing.

Finally, excess capacity in both producing countries and refineries is now providing the cushion previously expected from stocks. Variations in demand can be easily met now by altering the rate of refinery throughout. So why hold high volumes of stock? The sharp and partly unwarranted build-up of stocks in 1979–80, which continued in the early parts of 1981 and was followed later in 1981 and 1982 by a massive drawdown, have contributed to the destabilization of the oil market. In the first place prices shot up; in the second place demand fell below consumption requirements and this created considerable strains on the producing countries.

I believe that we have moved too radically from one extreme to another: insofar as stocks are concerned. There was too much panic in 1979–80 and there is too much complacency today. It is difficult to understand the total lack of concern about security of oil supplies manifested today by consciences when the Middle East is torn apart by two conflicts – the Arab-Israeli and the Iraq-Iran wars. Who can tell the complex and destabilizing repercussions of these two wars on the area? There are problems in Latin America too. Is it wise to ignore these issues and leave the consuming world unprotected by a stock cushion? The same complacency affected stocks in 1973 and in 1978. The absence of cushion induced panic when political accidents – the Arab-Israeli war of 1973 and the Iranian Revolution of 1979 – disrupted supplies. Alas, Governments and companies seem to have short memories and never seem to learn from the lessons of a very recent past.

Leaving now all these considerations aside, the fact remains that OPEC faces a depressed demand. Demand for OPEC oil supplies suffers from the double impact of a general decline in world demand and from the ability of non OPEC producers to maintain (or even increase) the volume of their exports even on slack markets.

Several years ago I pointed out that OPEC was being treated by the world as a residual supplier. This situation has considerable disadvantages both in the upswing, when OPEC countries are expected to meet any increase in world demand, and in the downswing when they are expected to absorb the whole reduction in world demand and hence on their income. OPEC must now devise a strategy which would enable it to retain control over the price of oil in a depressed market. Such a strategy must involve as a preliminary a reconsideration of the relationship with non-OPEC producing countries whose behaviour is putting OPEC in the disadvantageous position of the world residual supplier.

Non-OPEC producing countries are having it both ways. They take advantage of tight markets to raise prices to the maximum levels they can get. Recall that in 1979 the first official price increase was made by BNOC as early as January. OPEC countries began to follow suit much more timidly after a lag of 6 or 8 weeks. Everybody was after the highest price that one could get. The UK, Norway and Mexico were charging higher prices than the average OPEC countries without incurring any blame. They decided that their oil should be related to the high price African crudes, not to the Saudi marker which then provided a much lower benchmark. In 1981, when the tables began to turn, BNOC declared that its price will be linked to the Saudi marker no longer to the overpriced African crudes. The philosophy was clearly to maximise prices on the tight marker and to protect volumes when demand is slack. In 1982 the link with the Saudi marker was cut and BNOC, followed by Norway and subsequently by other producers undercut the OPEC price by three to four dollars. Why such a large discount when 50 cents would have done the trick and achieved their goal remains a mystery. But this is their business. What concerns OPEC is this propensity to undercut.

Non-OPEC countries cannot have it both ways. They should be reminded that if it pays them to undercut it is precisely because OPEC holds the price line. If their policies succeeded in undermining the OPEC price they would be engulfed in the collapse. They stand to lose substantially in revenues if prices drifted down as a result of their competitive underbidding.

OPEC should indicate very clearly to other producers that it is not prepared to tolerate indefinitely a situation in which it affords protection to those who systematically put its protection into jeopardy. The message should be: hold the price line or face the risks of a price collapse. Producers have common interests whether they belong to OPEC or not. A sharper perception of this common interest could save us all from disasters brought about by shortsighted competitive behaviour. Price harmonization would distribute the reduction in demand more equitably between producers both within and outside OPEC. Strains are aggravated when the distribution is unequal, when a group of producers have to carry the whole burden in their role of residual suppliers. Non-OPEC producers are unwise to shift all the burden on OPEC. They should remember that they do not have the power to fix the price of oil, that they have benefited and are benefiting from this power as exercised by OPEC, that they will not inherit this power which will pass onto the market should OPEC lose its grip, and that their current policies would create a situation in which OPEC's ability to regulate prices would weaken to the detriment of these same non-OPEC producers.

In a weak market OPEC also needs to define very clearly its price and production policies. In my lecture to this Seminar last year I emphasised the importance of an orderly price structure. The same message should be stressed again, with even greater urgency today. A disorderly price structure weakens OPEC in a slack market. The principles of an orderly price structure are simple:

(a)   a uniform price for crudes of similar kinds
(b)   price differentials which reflect correctly relative market values of different varieties of crude
(c)   no discounts or other under-the-counter arrangements which amount to price-cutting.

This could be underpinned by a production programme. There are well known difficulties in agreeing on a collective production policy. A further difficulty is that the programme should be sufficiently firm to signal to the market that producers are serious in their interest to hold prices, but also sufficiently flexible to enable them to respond to changes in demand conditions. But the difficulties are not inseparable. OPEC may learn to overcome them. After all its experience of production programming is very short dating from March this year. An OPEC cannot turn to precedents since it concerned itself almost exclusively during the twenty two years of its existence with prices.

Yet one can learn ones lesson from the past. The recipe for success is simplicity. OPEC succeeded on prices because it stuck to a very simple device: to define a price and hold it firmly in the interval between meetings of the Conference of OPEC Ministers. A production programme based on a simple formula could do the trick if members are prepared to implement it firmly, come what may, in the interval between stated meetings.

There is no doubt that a prolonged period of decreased demand would present a challenge to OPEC. But OPEC has proved throughout its history that it is capable of strong solidarity in times of emergency and crisis. Differences of objectives and interests, which naturally exist between countries with such diverse economic and political circumstances, are set aside whenever adverse market conditions seem to threaten OPEC's fundamental interest. We had a clear manifestation of this behaviour in the Vienna meeting of March 1982.

My belief is that no OPEC member wants to wreck the boat even though some members may on occasions deviate temporarily from the agreed line. Everbody stands to lose too much from a free-for-all. Oil-consumers and oil companies should not underestimate OPEC's determination to hold the price line and to retain the role of administrator of the price of crude oil.

But OPEC countries should not also over estimate the resilience of the system. There is danger of an accident which nobody wants to occur if members deviate too much from the agreed line. Temporary exceptions can be tolerated now and then, but a slow drift away from the common policy could suddenly snowball and put all OPEC's hard-won achievements in jeopardy. It is better to play it safe now – than to be sorry later.

In concluding, I would like to suggest that the difficulties faced today by the oil market, however important for producers and consumers of oil, are but one aspect of much more complex problems besetting the world. I have in mind the world economic crisis which has old roots that can be traced to earlier decades.

The solution of this crisis presents a challenge which Governments, economists and international institutions seem unable to solve. The economic troubles are coupled with political conflicts which may threaten the stability of the international system. These major issues deserve most of our attention and concerted efforts. The problem of oil cannot be divorced from the general environment.

# APPENDIX OF TABLES
# AND GRAPHS

ANNEX TO CHAPTER 1

Fig. 1 Exploration Expenditure (in Millions of US Dollars)

| Country | 1961 | 1962 | 1963 | 1964 | 1965 | 1966 | 1967 | 1968 | 1969 | 1970 |
|---|---|---|---|---|---|---|---|---|---|---|
| Middle East | 30 | 25 | 30 | 30 | 35 | 50 | 50 | 50 | 50 | 50 |
| Europe | 40 | 25 | 35 | 90 | 150 | 75 | 100 | 125 | 125 | 100 |
| U.S.A. | 600 | 575 | 600 | 650 | 610 | 650 | 615 | 715 | 725 | 665 |

Source: Chase Manhattan. Capital Investment of the World Petroleum Industry.

Fig. 2 Net Income of the Major Oil Companies (in Millions of US Dollars)

| Year | BP | CFP | Shell | Esso | Gulf | Mobil | Stancal | Texaco | TOTAL |
|---|---|---|---|---|---|---|---|---|---|
| 1961 | 168 | 22 | 524 | 758 | 339 | 211 | 294 | 434 | 2,750 |
| 1962 | 197 | 28 | 573 | 841 | 340 | 242 | 313 | 482 | 3,016 |
| 1963 | 232 | 36 | 601 | 1,020 | 371 | 272 | 322 | 548 | 3,402 |
| 1964 | 231 | 36 | 588 | 1,051 | 395 | 294 | 345 | 577 | 3,517 |
| 1965 | 225 | 42 | 628 | 1,036 | 427 | 320 | 352 | 637 | 3,667 |
| 1966 | 222 | 51 | 662 | 1,091 | 505 | 356 | 386 | 710 | 3,983 |
| 1967 | 180 | 56 | 732 | 1,195 | 578 | 385 | 409 | 754 | 4,289 |
| 1968 | 243 | 63 | 865 | 1,277 | 626 | 431 | 452 | 836 | 4,793 |
| 1969 | 232 | 73 | 946 | 1,243 | 611 | 457 | 454 | 770 | 4,786 |
| 1970 | 218 | 73 | 880 | 1,310 | 550 | 483 | 455 | 822 | 4,791 |

Source: OPEC Annual Statistical Bulletin.

*Fig. 3* How the Oil Barrel is shared (in US Dollars per barrel)

| | 1961 $ | 1961 % | 1970 $ | 1970 % | 1971 $ | 1971 % | 1972 $ | 1972 % | 1973 $ | 1973 % | 1974 $ | 1974 % | 1975 $ | 1975 % |
|---|---|---|---|---|---|---|---|---|---|---|---|---|---|---|
| Revenue of Producing Govt. | 0.76 | 6 | 0.86 | 6 | 1.35 | 8 | 1.60 | 9 | 2.30 | 11 | 9.10 | 34 | 10.10 | 30 |
| Taxes by Consuming Govt. in Europe** | 7.10 | 52 | 8.30 | 57 | 8.70 | 55 | 9.40 | 55 | 11.40 | 56 | 10.30 | 39 | 14.90 | 45 |
| Company Margins & Various Cost Elements | 5.70 | 42 | 5.30 | 37 | 5.90 | 37 | 6.00 | 35 | 6.80 | 33 | 7.30 | 27 | 8.20 | 25 |
| Total Weighted Average | 13.60 | 100 | 14.50 | 100 | 15.90 | 100 | 17.00 | 100 | 20.50 | 100 | 26.70 | 100 | 33.20 | 100 |

* On the basis of the Marker Crude.

**Calculated on Weighted Average of Products Consumption Patterns.

Source: Derived from product prices and taxes published in *"Petroleum Times"*.

ANNEX TO CHAPTER 3

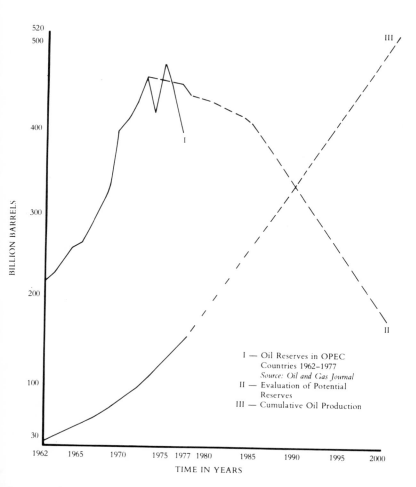

*Fig. 4* Crude Oil Reserves and Cumulative Production (in billions of barrels)

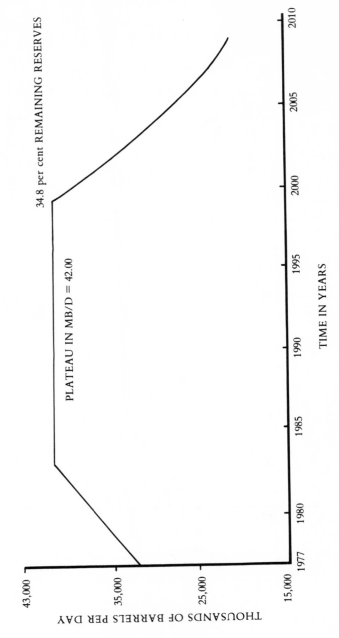

*Fig. 5* Total OPEC Production (in thousands of barrels per day)

# ABBREVIATIONS

Abbreviations employed in this volume

| | |
|---|---|
| ASEAN | Association of South East Asian Nations |
| BNOC | British National Oil Corporation |
| CIA | Central Intelligence Agency (US) |
| CIEC | Conference of International Economic Co-operation (Paris Dialogue) |
| COMECON | Council for Mutual Economic Assistance |
| EEC | European Economic Community |
| IEA | International Energy Agency |
| MITI | Ministry of International Trade and Industry (Japan) |
| OAC | Organization of African Countries – now changed to OAU, the Organisation of African Unity |
| OAPEC | Organization of Arab Petroleum Exporting Countries |
| OECD | Organization for Economic Co-operation and Development |
| OLADE | Organization of Latin American Energy |
| UNCTAD | United Nations Conference for Trade and Development |

# INDEX

Abu Dhabi
  decline in oil exports (1973–4), 111
  differentials, 123, 124–5
  gas exports, 101
Abu Dhabi meeting (Dec. 1981), 133
Adelman, Professor (*The World Petroleum
  Market*), 17
agriculture: development in Qatar, 103
aims of OPEC, 29, 50–3, 67
Alaska, 20, 77, 86
  gas exports, 101, 102
Algeria
  NOC, 66, 71
  gas exports, 101
alternatives *see under* energy
Arab-Israeli war (1973): embargo and
    price revolution, 82, 95, 122, 136
"arms-length transactions", 21, 22, 26
Attiga, Dr, quoted, 51–2

"brain drain" from Opec countries, 68
Brazil: Petrobras, 66, 72
British National Oil Co. (BNOC):
    price rise (1979), 137
Brunei: gas exports, 101

Carter, Pres. Jimmy: oil policy 83–4,
    87, 89–90
capital *see* finance
CIEC (Paris), 12–13, 50, 53, 54, 63
Clemenceau, Georges, on oil, 94
Club of Rome: warnings, 34
coal
  (1800–1972), 33–4
  gasification, 105
  in oil-consuming countries, 41
  substitution for oil, 134
  US development, 85
Comecon: gas reserves, 101
Committee on long-term strategies,
    113, 118

companies, independent
  competition, 20, 21
  co-operation with, 22, 26
companies (Majors)
  history, 81–2, 92–3
  as "Controlling Power", 18, 70–1,
    81–2, 108
  OPEC shares power (from 1960)
    with, 22–3
  and OPEC aims (1977), 29
  Oil Agreements (concessions), 18,
    66
  and OPEC negotiations on royalties
    (1964–    ), 49
  Tehran negotiations (1971), 29, 49
  profits, 20, 21, 22
  short-term interests of, 8, 108
concessions (oil agreements), 18, 66
conservation
  defined, 84
  laws (1960s), 23
  implementing of policy, 45
  required aims of consumer countries,
    40
  and fall in world consumption, 126,
    134
  and future production, 75
  and pollution, 38
  waste avoidance, 41–2
  alternative end uses, 25–7
  transition to alternative sources, 5,
    10–11, 36–9, 57, 62, 110
  gas flaring, 12, 43, 69, 77, 87
consumption
  oil: future indigenous, 75
  falling world (causes), 129–30, 134–5
  gas: future indigenous, 76
Controlling Power *see under* companies
currency variations
  effect on revenue, 45–6, 94, 112
Curzon, George, Lord, on oil, 94–5

# ABOUT THE AUTHOR

Mr. Ali M. Jaidah has been the Managing Director and Member of the Board of Directors of the Qatar General Petroleum Company since 1979.

His educational background includes degrees in Economics and Petroleum Economics from London University. Mr. Jaidah was Head of the Economics Division in the Department of Petroleum Affairs in the Qatar Ministry of Finance and Petroleum (1966–71), and the Director of Petroleum Affairs in the same Ministry (1971–76). He was Qatar Governor for OPEC and a Member of the Executive Office of OAPEC until 1976. He served from January 1977 to December 1978 as Secretary General of OPEC.

Concurrently with these activities, he actively participated in all negotiations relating to the oil industry in and outside Qatar, including those relating to the take-over of the equity of the Qatar Petroleum Company and the Shell Company of Qatar Ltd. He also headed Qatari delegations to OPEC and OAPEC, and participated in various other capacities in other international petroleum conferences.

Mr. Jaidah was awarded a decoration by the Federal President of Austria in recognition of his work as former Secretary General of OPEC: a silver medallion inscribed by the symbols of the nine States of the Federation, with a citation signed by the Federal President, Dr. Kirsch Lager.